MONSOON ECOLOGIES

MONSOON ECOLOGIES

Irrigation, Agriculture and Settlement Patterns in Rajasthan during the Pre-Colonial Period

MAYANK KUMAR

MANOHAR
2013

First published 2013

ISBN 978-93-5098-024-8

Published by
Ajay Kumar Jain *for*
Manohar Publishers & Distributors
4753/23 Ansari Road, Daryaganj,
New Delhi 110 002

Printed at
Salasar Imaging Systems
Delhi 110 035

In the memory of *Amma*
dedicated to *Papa*
and *Panditji*

Contents

THE RAIN-CATCHERS

Illustrations

MAP

PLATES

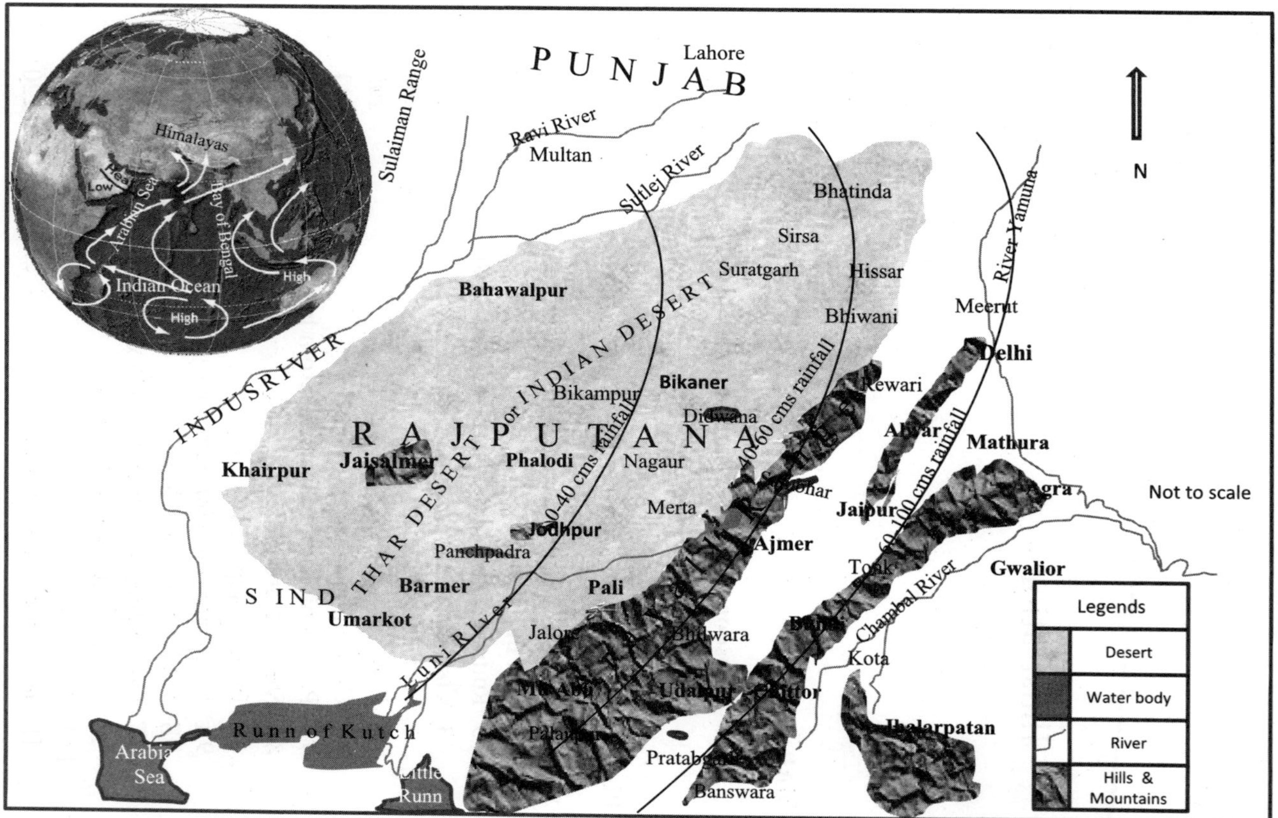

MAP I: MONSOON CURRENTS, SETTLEMENTS, PHYSICAL FEATURES AND RAINFALL PATTERNS

Preface

More than a hundred years have passed since the first modern irrigation system was commissioned under the aegis of the British rule in India. The then British government had incurred huge investments in this area. However, the post independence era has witnessed much heavier investments in the field of irrigation. A substantial part of these investments have been channelized towards canals, tube wells and wells. What would be the importance of different 'traditional systems' of irrigation, e.g. *Beri*, *Bahlo*, *Khareen,* and *Kundi* under the current Indian scenario?

It was with the pioneering field work undertaken by the Centre for Science and Environment that a comprehensive documentation of traditional water harvesting system was accomplished. The title of the work, *Dying Wisdom*, conveys a lot about the deteriorating condition of these traditional systems of irrigation in India. In fact this study was a timely reminder not only to the policy makers but also to the academic world in general and historians in particular towards a prominent blind-spot in their researches. During early years of the decade of 1990 a seminal work was published by Madhav Gadgil and Ramchandra Guha. This work placed environment and environmental concerns at least in the domain of historical investigation in India.

Against the backdrop of these developments my present work may be said to be the outcome of two important considerations:

1. Most of the modern irrigation systems are energy intensive and usually under the control of state and,
2. Recent researches in the history of climate have made it clear that variability is an inherent feature in the history of climate and will continue to be an important factor determining our response to changes in climate. Human societies have adapted to these changes in the past and need to adapt in future as well.

Broadly speaking, these two considerations prompted me on the one hand to investigate the history of monsoon in India and locate variability, if possible, and on the other to understand the processes of adaptation primarily in terms of negotiations with the erratic behaviour of monsoon in India. Further, it is expected that these adaptation strategies might offer us insights to develop systems to respond to climate variability. In the light of these considerations, Rajasthan emerged as a natural choice, more so, due to availability of rich historical evidences; primarily archival documentation offering gleanings to the process of adaptations at the level of primary agrarian production. Moreover, agrarian studies on Medieval and early-Medieval India in general and on the region in particular offered a firm background to locate the 'environment' in the historical processes. Further, my doctoral thesis helped me understand the human–nature interactions in the semi-arid and arid landscape of Rajasthan and facilitated in my endeavour to investigate social responses in years of variability as well.

I acknowledge my sincere gratitude to my supervisor Prof. Dilbagh Singh for his vision and trust in enrolling me under his supervision for a relatively new field in Indian history. He not only welcomed this idea of investigation of human–nature interaction but continued to shape the research with his deep insight of the agrarian economy and tremendous knowledge of the archival records available for Rajasthan. I must share his insistence on field-visit to the region I was working on.

It is difficult to express my gratitude to Guddu da, Pankaj da and Ajit da. They have been the continuous source of inspiration, guidance and motivation. Field-trips became memorable due to company of Bhabhi and Sandy. Trips were hassle free as Basu took good care of Dooj. The excitement of Sadhu and Aseem to accompany me during field-trips never diminished though could never materialize. Mona's uncle and his family have been a gracious host in a very remote village. Visit to Kakuni Mata and archaeological remains of Dalhan were made possible by Manish and Mona's ingenuity to organize it with the help of her cousin. Continuous support from Guddi Bhabhi, Nirat Bhabhi, and Purnima Bhabhi always encouraged me during times of low esteem.

The blessings of my in-laws, Jijaji, and Geeta have continuously supported my endeavour. Amit da and Jayati have been a regular source of ideas not only for this book but my research interests as well. Photography tips of Neeraj and editing tips of O.P. have been as valuable as our friendship. Shashank and Nilanjan deserve thanks for not only the continuous encouragement but also for making me acquaint with latest publication on related themes. More than the insightful remarks, Rohan D'Souza, Bharatji, Neerjaji, Rajesh, Khagendra, and Nagendra, have shared moments of despair and have always encouraged me to accomplish the work.

I acknowledge with happy memory many other debts: Prof. Mahesh Ranagarajan, has always pushed me to go beyond the conventional confines of the discipline and region. It is difficult to remember even a single meeting with him when he has not suggested new readings for my research often from distant land/region. I must thank him for unwearyingly correcting my formulations and time and again suggesting new ways of looking at the evidences available. Prof. G.S.L. Devra had always something new to offer on the nuances of human–nature interactions. I also wish to thank to my seniors and teachers, Profs. Harbans Mukhia, Bhupender Yadav, Yogesh Sharma, Sunita Zaidi, Inayet Zaidi, B.L. Bhadani, Yogendra Singh, Rajat Dutta, Mihir Deb, Drs. Biswamoy Pati, Meena Bhargava, R.P. Bahuguna and Suraj Bhan for their valuable comments, discussion, and guidance.

My colleagues in the college especially, Tulikaji, Prashantji, and Thakur Sahab, have also been a constant source of encouragement for doing this work.

First hand acquaintance with Rajasthan was made possible by the financial support initially provided by UGC and later on by ICHR. Association with Decision Centre for a Desert City, as a Fulbright Fellow, made me realize that though monsoon has global signature but human interactions differ from region to region as interactions with climate are social constructions and are usually defined in terms of some sort of socio-political-economic-cultural confines. Mike with his GIS tools and Andrew with his eagerness to compute the data forced me to locate the data beyond the confines of Indian Metrological Department. I owe big thanks to Mike, Andrew, Sally, Bill, Darren, David, Dave, and Patricia for elaborately and patiently discussing the peculiarity and nuances of monsoon in south-west USA. Interactions with Anne, Megha, Sadia and Sammit have been immensely useful as they reminded me time and again not to ignore the diversity between Rajasthan and other regions of Indian peninsula, like, Deccan, Marathawada, Sind and Western Ghats. Later on, my two years' stint with Nehru Memorial Museum & Library as fellow helped me integrate the global with regional/local. I am thankful to both the institutions for offering me immense opportunities to acquaint myself with the latest in the field. Library staff has been very supportive and deserve a special thanks. Thanks are also due to all the scholars whose works have been used by me in my work.

I must also thank anonymous reviewer of the manuscript for his comments and suggestions. Sh. R.C. Jain of Manohar has shown keen interest in the manuscript and very graciously accepted my suggestion for retaining more than ten pictures. Thanks.

Last but not the least, I must acknowledge that though it was difficult for Dooj to understand the significance of a monograph but he became well aware that it takes time, so rarely complained about me for not devoting sufficient time to him. Endurance of Raj cannot be described in words so let it be unsaid as has been my habit. Simple thanks will be insufficient.

I am, however, alone accountable for the shortcomings of the work.

Delhi MAYANK KUMAR
21 May 2013

Introduction

The Indian subcontinent offers a unique opportunity to researchers to examine the human–nature interactions in different ecological niche within the larger umbrella of monsoon ecology. The centrality of the monsoon for the emergence, growth and endurance of human settlements in different parts of the Indian subcontinent needs to be reiterated, revisited and contextualized with respect to variances/differences in the human negotiations in different parts of the subcontinent. Perhaps it is the nature of human-ecology interactions which have given a peculiar identity to the region and its people in Rajasthan. It will not be out of place to mention that the history and culture of Rajasthan have largely been portrayed as the history of bravery, chivalry and sturdiness of Rajputs, surviving in the harsh climate conditions, particularly in the arid zone of the region. The question arises, whether the history of Rajasthan should necessarily be read and understood as the history of Rajputs and its desert only or there can be other equally important but different readings of historical pasts of the region. Moreover, this dominant characterization has been, to some extent, responsible for the neglect of the significance of inherent ecological and cultural diversity of the region. Whatever examination of the ecological diversity has been undertaken is usually in terms of regional polities, thus subjugating the importance of ecological settings to the dynasty-centric narratives. Nevertheless, the significance of monsoon and the vagaries associated with it are important unifying factors. In the tropical landscape of the Indian subcontinent it is the monsoon which often offers us the vantage point to examine the interactions between its people beyond the regular categories of region, regional polities, and temporal divisions.

The monsoon forms the backbone of Indian agriculture in the deltaic zones of Bengal, in the cold landscapes of the Himalayan slopes, in the fertile plains of the Indo–Gangetic plains and also in the semi-arid and arid landscapes of Rajasthan and Sind. The predominant significance of monsoon rains in human settlements in the ecological nichés cutting across different physiological regions of India cannot be underestimated. The arid and semi-arid landscape of the Thar Desert and the adjoining areas on both sides of present-day India–Pakistan border is an ideal landscape to carry out such an examination. The significance of monsoon ecologies to sustain human settlements becomes most prominently visible at these points where human ingenuity is tested and a newer set of adaptations are shaped. The role of climate in general and the monsoon in particular, in the evolution and growth of human settlements has been scarcely touched upon. Therefore, a case study is being attempted to offer more nuanced readings of human–nature interactions. Integrating data from palaeoecology, archaeology and history, I have treated the environmental, social and political transformations in this region over the last one-and-a-half millennia, focusing on the growing importance of monsoon rains during early medieval times when, I argue, many of the strategies available to contemporary societies were tested and led to a different set of adaptations.

Coupled with the diversity of landscapes covered largely by the singular climatic phenomenon of the monsoons, the Indian subcontinent also offers a very long and deep continuity of human settlements. The continuity of settlements in different regions carry with them a history of adaptation and accommodation; at times apparent and most of the times not so apparent.[1] At another

[1] Sumit Guha suggests that the human–nature relationships in early India were marked by fluidity and a state of flux. There was a longer process of continuous interaction among communities of several ecological nichés like riverine plains, deserts, forests, mountains. Guha has argued that pastoralists

plain, the histories of these adaptations and accommodations are also manifestations of the climatic variability along with the changes in the socio-economic-political settings. Uninterrupted habitations at times impede the examination of the specific cases of accommodations and adaptations. Continuity of human settlements generally hampers the historical investigations as it restricts the possibilities of archaeological explorations, but by the same token offers us other sources/evidences to explore deep into the past processes. Long surviving folk traditions along with a plethora of information in the form of rituals and cultural practices apart from the written records, both literary and archival, of past societies presents before us an excellent repository of insights. Long historical traditions put forward before the historian a mine of possibilities to examine human–nature interactions.

It is interesting to note that the longevity of settlements apparently presents an image of the singularity of adaptation/s and thereby restricts the nuances of adaptations and accommodations.[2] This has often given rise to a romantic image of societies of the Orient[3] and has been used as an excuse to justify the colonial intervention.[4] Lack of in-depth historical investigations of processes of adaptations evolved during the periods of climatic variability

---

and agriculturists were dynamic categories. Herdsmen traversed both the forests and the ploughed land. Forests provided not only resources and revenue but also refuge to rebels. Sumit Guha, *Environment and Ethnicity in India, 1200–1991*, Cambridge University Press, Cambridge, 1999, p. 7.

[2] *Karl Marx, Articles on India*, Peoples' Publishing House, Bombay, 1943; Iqbal Husain (ed.), *Karl Marx on India*, Tulika, Delhi, 2006.

[3] James Tod, *Annals and Antiquities of Rajasthan*, vols. I and II, Rupa & Company, Delhi, 1997; Madhav Gadgil and Ramchandra Guha, *This Fissured Land: An Ecological History of India*, Oxford University Press, Delhi, 1992, pp. 1–108; B.H. Baden Powell, *The Indian Village Community with Special Reference to the Physical, Ethnographic and Historical Conditions of India*, Cosmo Publications, Delhi, 1972.

[4] Vincent A. Smith, *Oxford History of India* (1919) 3rd revised and ed. Mortimer Wheeler, A.L. Basham and T.G. Percival Spear, Oxford University Press, Oxford, 1958. 'Not many would today defend Smith's openly imperial stance. . . . Embedded in it are certain assumptions that were not his alone. Nor were they confined to the upholders of the British Raj. Smith's Indology belonged to a much larger movement of world dominance, one that has changed a great

coupled with the success of adaptations which could fortunately and/or unfortunately skip long-term dislocations, perhaps with the exception of the Harappan Civilization, has only contributed to the perpetuation of this image.[5] Even the early writing which can perhaps be termed as nascent environmental history often written by historical geographers, initially developed to trace the role of climate in the decline of the ancient civilizations.[6] None the less, despite the well-established fact that the climate has witnessed variability in the past, the continuation of settlements in the Indian subcontinent has only reinforced the image that Indian society has not witnessed any change since ancient times till the imposition of colonial rule.

Since the singular significance of climate and landscape in the emergence and growth of settlements has usually been examined

---

deal since his time but is still very much with us.' Ronald Inden, *Imagining India*, Basil Blackwell, Oxford, 1990, p. 9.

[5] The debate over the decline of Harappan Civilization has attracted scholarly attention perhaps due to other reasons also. The question of Sarasvati and the ancient origins of Aryans have provided an impetus to such studies and it is a flourishing area of investigation. B.P. Radhakrishna and S.S. Merh (eds.), *Vedic Sarasvati: Evolutionary History of a Lost River of Northwestern India*, Memoir 42, Geological Survey of India, Bangalore, 1999; Shereen Ratnagar, *Understanding Harappa: Civilization in the Greater Indus Valley*, Tulika, Delhi, 2001; Rajesh Kochhar, *The Vedic People: Their History and Geography*, Orient Longman, Hyderabad, 2000; Edwin F. Brynt and Laurie L. Patton (eds), *The Indo-Aryan Controversy: Evidences and Inferences in Indian History*, Routledge, London and New York, 2005; Jane R. McIntosh, *The Ancient Indus Valley: New Perspectives*, ABC-Clio, Inc, Santa Barbara, 2008; R.N. Nandi, *Ideology and Environment: Situating the Origins of Vedic Culture*, Aakar Books, Delhi, 2009; Thomas R. Trautmann, *The Clash of Chronologies: Ancient India in the Modern World*, Yoda Press, New Delhi, 2009.

[6] For the region of present study among the early writers one needs to look at the following works: Major H.G. Raverty, *The Mihran of Sind and its Tributaries*, Sang-E-Meel Publications, Lahore, 1979; Surgeon-Major C.F. Oldham, 'The Saraswati and the Lost River of the Indian Desert', *Journal of Royal Society*, 1893, pp. 49–76; Richard Grove and Vinita Damodaran, 'Imperialism, Intellectual Networks, and Environmental Change: Origins and Evolution of Global Environmental History, 1676-2000', Part I, *Economic and Political Weekly*, vol. XLI, no. 41, 14 October 2006, pp. 4345–54; Part II, *Economic and Political Weekly*, vol. XLI, no. 42, 21 October 2006, pp. 4497–4505.

at the regional/national levels, both the local variance in itself and changes over a long temporal frame have been a major casualty. Indian civilization has usually been attributed an antiquity generally giving the impression of a long-term continuum[7] often concealing the different temporal locations for different settlements within a larger region. The application of antiquity has unfortunately often obscured the examination of the processes of changes and adaptations. Even the settlements which emerged later, perhaps in responses to climate variability, have mostly been attributed similar 'antiquity' thereby negating the specificity and denying the examination of each case in its specific historical–ecological context.[8] This perhaps explains the ever-growing debate on the early medieval social formation of the Indian subcontinent.[9] The contestations and conflicts over the nature of historical processes is a clear manifestation of the disagreement over the 'cause/s'. Explanations in terms of dominant historiography whereby

[7] In their seminal work, *This Fissured Land*, even Guha and Gadgil have argued that the Indian caste system created a prudent and harmonious utilization of resources as separate castes monopolized economic activities and these castes had their own ecological niche's. Although this work raised serious questions regarding resource use, however, the notions of immobility and self-sufficiency are very much implicit in the arguments put forth.

[8] Sumit Guha challenges the prevalent practice of attributing an image of immobility and self-sufficiency on communities. Guha, *Environment and Ethnicity in India*, p. 43.

[9] B.D. Chattopadhyaya, *Aspects of Rural Settlements and Rural Society in Early Medieval India*, K.P. Bagchi & Company, Calcutta and New Delhi, 1990; *The Making of Early Medieval India*, Oxford University Press, New Delhi, 1994; Harbans Mukhia, 'Was There Feudalism in Indian History?', *Proceedings of Indian History Congress*, 40th Session, Waltair, 1979, pp. 229–80; R.S. Sharma, *Indian Feudalism: c. 300–1200*, Calcutta, 1965, 2nd edn., Macmillan India, New Delhi, 1980; *Urban Decay in India (c. AD 300–1000)*, Munshiram Manoharlal, Delhi, 1987; *Social Change in Early Medieval India (c. AD 500–1200)*, Peoples' Publishing House, Delhi, 1969; M.K. Dhavalikar, 'The Golden Age and After: Perspectives in Historical Archaeology', *Proceedings of Indian History Congress*, 60th Session, Calicut, 1999, pp. 1–18; Cynthia Talbot, *Pre-Colonial India in Practice: Society, Region and Identity in Medieval Andhra*, Oxford University Press, New Delhi, 2001; Burton Stein, *Peasant, State and Society in Medieval South India*, Oxford University Press, Delhi, 1980.

attributing primacy to the social factor is now being consistently challenged by extending the argument that climatic factors are also important players. Unfortunately these initial writings tend to restrict the sphere of environmental factors to irrigation and agriculture. K. Sivaramakrishnan and Ismael Vaccaro have aptly argued against such a narrow interpretation of environmental factors. They argue:

> . . . to talk of nature, is not merely to talk of entities and domains, it is always also to imagine and describe relationships and connections. It is, however, hard to capture complex relationships between human–cultural and biophysical processes quite so easily, in any instrumental language of design, action, outcome and aftermath.[10]

To capture the interactions 'in any instrumental language of design, action, outcome and aftermath' is a very difficult task, more so because we are trying to capture the interactions between two very dynamic entities, that is, climate and the human capacity to adapt. There is no need to emphasize the fact that the 'environment' is not merely a static entity but has a living existence. It changes over time and is interconnected with wider space and hence demands a careful understanding of this process of change, thus necessitating expansion of the domain of historical investigations itself. Lucien Febvre says that:

> in the explanation of these very complex facts which have been subject to varied circumstances of time and places, geographical analysis as well as that of ethnic and historical influences ought to have its share. The exclusive use of one method of interpretation would not satisfy a mind devoted to truth and to a system.[11]

He further stresses that:

> As matter of fact, it is only a question of possibilities. Again if the influence of the local physical environment is undeniable, does that not mean that

[10] K. Sivaramakrishnan and Ismael Vaccaro, 'Introduction: Postindustrial Natures: Hyper-mobility and Place-attachments', *Social Anthropology*, vol. XIV., no. 3, 2006, pp. 301–17, here p. 303.

[11] Lucien Febvre, *A Geographical Introduction to History*, tr. E.G. Mountford and J.H. Paxton, Kegan Paul, Trubner & Co. Ltd., London, 1925, p. 49.

a population of emigrants has conceived on another soil and in another climate, sometimes even the structure of the village? Can it not be possible that the newcomers built and arranged their dwelling in a pattern familiar to their fatherland? May not this type have been modified if experience did not allow it to be kept unchanged?[12]

Similarly, the other component, that is, the humans in the nature–human interaction also exhibit a great amount of diversity. At times, the variance in social structure is closely related to the difference in geographic conditions. It is manifested in the pattern of land use, irrigation, agrarian practices, settlement patterns, economy subsistence patterns, conservation practices, etc. The geographical specificities of the various parts of the region help us to broadly understand how ecological features have influenced and shaped this social variance. Even more dynamic are the emergence of perceptions, ethics, myths and other structures of meaning and the process through which they become part of an individual or groups' dialogue with nature in general and climate in particular. The response manifests itself in a cultural diversity of thoughts, ethics and social practices. Since all human societies are continuously engaged in defining and constructing the world around them, the question of social beliefs also assumes significance. This endeavour is also important as human adaptability to any climate depends in part on the society's perception/s of climate and the changes occurring thereof. In their attempt to capture the dynamism of human–nature relationships, David Arnold and Ramchandra Guha have suggested that 'knowledge of environment as expressed in the popular perceptions and experience of folk traditions and religious beliefs are more familiar to anthropologists than historians'.[13] Gunther-Dietz Sontheimer's contribution to the study of the cult of Mallana/Mailara in a larger contiguous zone spread over Andhra, Karnataka and Maharashtra offers important insights

[12] Ibid., p. 47.

[13] David Arnold and Ramchandra Guha, 'Introduction: Themes and Issues in the Environmental History of South Asia', in David Arnold and Ramchandra Guha (eds.), *Nature, Culture and Imperialism: Essays on the Environmental History of South Asia*, Oxford University Press, Delhi, 1995, pp. 4–5.

into society located on the fringes of agriculture.[14] Similarly, Ann Feldhaus, while examining the *nadi mahatmya* to understand the significance of the river in the semi-arid landscape of Maharashtra, has elaborated upon the social perceptions of climate in the region and how gender has been an important category in historical investigations.[15]

Not to undermine the significance of but at the same time arguing against the singularity of the agrarian landscape in historical investigations, David Ludden suggests that 'farming landscapes are therefore not defined primarily by their physical or environmental qualities, but rather by the long-term interaction of geography, culture, technology and social power'.[16] Changes in the landscape due to newer sets of negotiations with climatic variability have been documented by Kathleen Morrison for southern India in general and the Vijayanagara region in particular. She suggests that one witnesses continuous interactions between given landscape and social action. This interaction over a long period of time modifies and changes both—landscape as well as ways and means of social interactions.[17] Based on his work on the study of tank irrigation in the Ramnad–Sivaganga region of southern Tamil Nadu, Mosse looks historically at the interactions of water, landscape, culture and power as they have unfolded over hundreds of years in a region of scarcity and notices the inherent complexities in the interactions.[18]

[14] Gunther-Dietz Sontheimer, 'God as King For All: The Sanskrit Mallari Mahatmya and its Context', in Anne Feldhaus, Aditya Malik and Heidrun Bruckner (eds.), *King of Hunters, Warriors and Shepherds*, Manohar, Delhi, 1997, pp. 213–14.

[15] Anne Feldhaus, *Water and Womanhood: Religious Meanings of Rivers in Maharashtra*, Oxford University Press, New York & Delhi, 1995.

[16] David Ludden, *An Agrarian History of South Asia: The New Cambridge History of India*, vol. IV. 4, Cambridge University Press, Cambridge, 1999, p. 49.

[17] Kathleen D. Morrison, 'Environmental History, the Spice Trade, and the State in South India', in Mahesh Rangarajan and K. Sivaramakrishnan (eds.), *India's Environmental History*, vol. I: *From Ancient Times to the Colonial Period*, Permanent Black, Ranikhet, 2012, pp. 296-326; Kathleen D. Morrison and L.L. Junker (eds.), *Forager-Traders in South and Southeast Asia: Long-Term Histories*, Cambridge University Press, Cambridge, 2002.

[18] David Mosse, *The Rule of Water: Statecraft, and Collective Action in South India*, Oxford University Press, Delhi, 2003.

Most of these studies have documented social interactions to a given landscape with more or less constant climatic conditions, often reducing the impact of variability in the climate to nought. This tendency has survived for long due to greater emphasis on the larger regional picture often undermining the variance visible at village/local levels. Thus, it will not be wrong to point out that the significance of climatic variance resulting in adaptations at local/regional level without disturbing the larger social formations or political edifice was not even considered to be worthy of historical investigations. Climatic variability, as a factor influencing the growth and continuance of settlement patterns has emerged as a factor very recently, especially in India. More so because global dimensions of monsoon patterns became visible to humans very recently and since then there have been continuous efforts by scientists to unravel the mysteries of the monsoon. El Nino/La Nino-Southern Oscillation effects and their mutuality have very recently become visible to the academic world. The mutuality of these effects in influencing the patterns of climate in the areas dominated by monsoon rains has opened up new avenues to decipher the signature of climatic patterns and its impact on human settlements in different parts of the globe.[19] The global interconnections of climatic patterns have been relied upon, if possible to develop a comparative between two different civilizations located in two different landscapes but part of a larger global climatic pattern.[20]

Though in the recent past, there has been tremendous advance in our understanding of past climate primarily due to developments

[19] Richard H. Grove and John Chappell, 'El Nino Chronology and the History of Global Crises During the Little Ice Age,' in Richard H. Grove and John Chappell (eds.), *El Nino: History and Crisis Studies from Asia-Pacific Region*, The White Horse Press, Massachusetts, 2000.

[20] M.K. Dhavalikar, 'Green Imperialism: Monsoon in Antiquity and Human Response,' *Man and Environment*, vol. XXVI, no. 2, 2001, pp. 17–28; Ranabir Chakravarti, 'The Creation and Expansion of Settlements and Management of Hydraulic Resources in Ancient India,' in Richard Grove, Vinita Damodaran and Satpal Sangwan (eds.), *Nature and the Environment*, Oxford University Press, Delhi, 1998, pp. 87–105; 'Agricultural Technology in Early Medieval India (*c.* AD 500–1300)', *The Medieval History Journal*, vol. XI, no. 2, 2008, pp. 229–58.

in the field of palaeoclimatic studies,[21] yet the construction of past climates is inconclusive. Most of the formulations lack consensus among scientists.[22] Thus, intervention on the part of the historian is necessary as at times they can offer corroboratory evidence on the basis of a different set of data. As reported by Ladurie,

> . . . when meteorologists, geologists, and biologists wanted to find out about the climate of the eleventh and sixteenth centuries, they invited a dozen rural and economic historians to their meeting. And they entrusted to them the task of establishing most of the series for the periods concerned, the historians having brought along with them the continuous, annual, quantitative, homogeneous lists of figures which the scientists had need of. So the relations between the history of climate and the related disciplines involve a fruitful exchange, a constant flow of information in both directions.[23]

It makes sense as 'historical records have the advantage of delineating the social and economic impacts of extreme climatic events and passages, which physical and biological recorders of climate such as coral growth bands or tree ring series cannot do'.[24] It is easier said than done as verification by the historians is not an easy task in the absence of coherent historical records of past

[21] However, it is important to note that science/s which can map the historical process of changes in vegetation are still in a nascent stage and it will take a long time to conclusively prove either the process of change or the change itself solely in terms of mathematical scientific evidences. Hannah M. David, Paul J. Wood and Jonathan P. Sadler, 'Ecohydrology and Hydroecology: A "New Paradigm"?', *Hydrological Processes*, vol. XVIII, 2004, pp. 3439–45.

[22] Marco Madella and Dorian Q. Fuller, 'Palaeoecology and the Harappan Civilisation of South Asia: A Reconsideration', *Quaternary Science Reviews*, vol. XXV, 2006, pp. 1283–1301. One can also see limitations of new tools/disciplines in predicting the precipitation, Ulrich Von Rad, Michael Schaaf, Klaus H. Michels, H. Klaus, H. Schulz, H. Wolfgang Berger and Frank Sirocko, 'A 5000-Year Record of Climate Change in Varved Sediments from the Oxygen Minimum Zone off Pakistan, Northeastern Arabian Sea', *Quaternary Research* 5, 1999, pp. 39–53.

[23] Emmanuel Le Roy Ladurie, *Times of Feast, Times of Famine: A History of Climate Since the Year 1000*, tr. Barbara Bray, Doubleday & Company, Inc, New York, 1971, p. 21.

[24] Grove and Chappell, 'El Nino Chronology', p. 5.

societies. The task attains colossal proportions in the case of pre-historic and/or proto-historic societies when historians to a great extent rely solely on archaeological evidences. Corroboration of a palaeoclimatic construction of the human past with recorded history has its own limitations and has often been proved insurmountable. Moreover, it is not easy to offer a well-knit narrative of human–nature negotiations as sustenance patterns show tremendous heterogeneity along with often near total absence of any written evidence. The task becomes more complex as this narrative suggests that nature–human negotiation may not necessarily offer a linear progression and deviations should be seen as the strength of the society rather than as weakness or decline as argued in this monograph.

CLIMATE IN HISTORICAL WRITINGS

The early writings had very different orientations with respect to climate, as they often attempted to correlate the human social and physical characteristics to a specific climate.[25] It was generally believed that the study of the climate and its features can explain the broad characteristics of culture, religious practices, economic orientations, etc., along with physical attributes and mental make-up. This classical understanding of the climate had to a certain extent racial connotations as well and flourished under the banner of European imperialism. Mark Harrison in his preface to *Climate and Constitutions: Health, Race, Environment and British Imperialism in India, 1600–1850*, suggests that '[t]his book charts the rise, growth and eventual decline—though not extinction—of ideas of climatic determinism, as expressed in European writings on Indian subcontinent'.[26] An unintended by-product of such a correlation had been extensive documentation of the climate of different parts of the globe along with detailed descriptions of

[25] Abul Fazl, *Ain-i-Akbari*, vol. III, tr. Col. H.S. Jarrett, Aadiesh Book Depot, Delhi, 1978, pp. 50–3.

[26] Mark Harrison, *Climate and Constitutions: Health, Race, Environment and British Imperialism in India, 1600–1850*, Oxford University Press, Delhi, 1999, p. iii.

social-economic-political profiling. Richard Grove and Vinita Damodaran have rightly argued that this documentation initially enriched the discipline of 'historical geography' and one can also trace, 'the intellectual origins of environmental history as a self-conscious domain of enquiry . . . to the encounters of 17th and 18th century western Europeans with the startlingly unfamiliar environments of the tropics'.[27] Even the modern climatological histories have not shied away from examining the impact of climate and changes in the climate on human affairs.[28] Ladurie suggests that:

> The explanation lies, in the first place, in the attitude adopted by the first historians of climate, who instead of studying climate change in itself and for its own sake launched themselves into something quite different and highly dangerous, namely the climatic interpretation of human history. Elsworth Huntington, for example, did not really make an unprejudiced study of climatic fluctuations in Asia, but tried from the outset to account for the Mongol migrations in terms of climate.[29]

In general, history of climate has gradually refined its 'resolution' of data and is now more than ever better placed to pin-point its estimates more accurately. Climatalogical historians have broadened their concerns beyond traditional subjects of temperature and precipitation by addressing issues like circulation patterns, storm activities, etc., along with a process of record-linkage which connects scattered bits of historical evidences into coherent general patterns of climate change.[30] It is disturbing to note that

> . . . the disciplinary gulf between the social and biological sciences (*even physical*) has, to a great extent, obscured much of the complex

[27] Grove and Damodaran, 'Imperialism, Intellectual Networks'.

[28] H.H. Lamb, 'An Approach to the Study of the Development of Climate and its Impact in Human Affairs', in T.M.L. Wigley, M.J. Ingram and G. Farmer (eds.), *Climate and History: Studies in Past Climate and Their Impact on Man*, Cambridge University Press, Cambridge, 1981, pp. 289–90.

[29] Ladurie, *Times of Feast, Times of Famine*, p. 7.

[30] David Hackett Fischer, 'Climate and History: Priorities for Research', in Robert I. Rotberg and Theodore K. Rabb (eds.), *Climate and History: Studies in Interdisciplinary History*, Princeton University Press, Princeton, New Jersey, 1981, pp. 241–50.

imbrications between socio-politics and landscape history, with the result that human-land relationships, (*even human-climate*) if not simply ignored, are, conversely, typically seen as part of the essential 'nature' of social groups rather than as the contingent products of long-term, mutually constitutive actions and interactions.[31]

None the less, the examination of climate as a factor in historical investigation gained due recognition only with the efforts of the *Annales* school of history writings. In their endeavour to expand the domain and concerns of the historian, the pioneers attempted to write 'total history' which fortunately offered a great amount of recognition to the geographical factors. In their attempt to document the totality of the historical experience of an age, *Annales* historians even attributed the authority of agency to the physical world as well. Ladurie acknowledges that, 'It was in fact the history of agriculture that led me, by a logical and even inevitable transition to the history of climate.'[32] Thus, it is to be noted that unlike climatologists or historians of climate, *Annales* historians were not particularly investigating the role of climate; rather, by granting agency to climate they postulated that the role of physical context of any society needs to be integrated for a comprehensive examination of the historical processes. Further, deliberating on the significance of climatic history and future possibilities, he argued that,

. . . at a second stage in our research, in which climate is no longer looked at its own sake but 'as it is for us', as the ecology of man. Climatic history would then become ecological history, asking such questions as whether the fluctuations of climate—or to put it more modestly, the brief fluctuations of meteorology—have reacted on the human habitat; on

[31] Morrison, 'Environmental History, the Spice Trade', p. 298. The monsoon was recognized as an important area of investigation but studies were confined to the concerns of poets, philosophers, physicians, etc., and occasionally historians. See Jay S. Fein and Pamela L. Stephens (eds.), *Monsoon*, John Wiley & Sons, New York, Chichester, Brisbane, Toronto, Singapore, 1987. It was very late that National Book Trust published a monograph on monsoon; P.K. Das, *Indian Monsoon*, National Book Trust, New Delhi, 1968, rev. edn. 2002. In the same vein one can find the following interesting and relevant monograph: Liba Taub, *Ancient Meteorology*, Routledge, London and New York, 2003.

[32] Ladurie, *Times of Feast, Times of Famine*, p. 2.

harvests and thus on economy; on epidemics and diseases, and thus on demography.'[33]

It is not surprising to note that in general, historians found it more convenient to address questions like the role and impact of 'fluctuations of climate . . . on the human habitat; on harvests and thus on economy; on epidemics and diseases, and thus on demography.'[34] The reasons are not very difficult to decipher. Given the limitations of the tools and sources, it is not always possible and feasible to recreate the history of climate for different regions of the world. Nevertheless, that does not mean that the role of climate in historical investigation can be undermined. The role of climate as a factor in the emergence and decline of civilizations has been recognized for a fairly long period of time.[35] Similarly, catastrophic impacts of climatic changes/fluctuations have also generated a fair amount of writings. Although the recent past has seen a tremendous spurt in historical writings with climate and associated fluctuations at the centre,[36] in general it is the history of famines and other related issues which have offered greater agency to the climatic factors.[37] The significance of the physical context for historical

[33] Ibid., p. 22.

[34] Theodore K. Rabb, 'The Historian and the Climatologist', in Robert I. Rotberg and Theodore K. Rabb (eds.), *Climate and History: Studies in Interdisciplinary History*, Princeton University Press, Princeton, New Jersey, 1981, pp. 251–7.

[35] F.R. Allchin, *The Archaeology of Early Historic South Asia: The Emergence of Cities and State*, Cambridge University Press, Cambridge, 1995; V.N. Misra, 'Climate, a Factor in the Rise and Fall of the Indus Civilization: Evidence from Rajasthan and Beyond', in B.B. Lal and S.P. Gupta (eds.), *Frontiers of the Indus Civilizations*, Books and Books, Delhi, 1984, pp. 461–6, 473, 481–2; Gurdeep Singh, 'The Indus Valley Culture (Seen in the Context of Post-Glacial Climate and Ecological Studies in North-West India)', *Archaeology and Physical Anthropology in Oceania*, vol. VI, no. 2, pp. 177–89; D.P. Agrawal, *The Copper Bronze Age in India*, Munshiram Manoharlal, Delhi, 1971.

[36] Brian M. Fagan, *The Long Summer: How Climate Changed Civilisation*, Basic Books, New York, 2005; *The Great Warming: Climate Change and the Rise and Fall of Civilizations*, Bloomsbury Press, New York, 2008; David R. Montgomery, *Dirt: The Erosion of Civilizations*, University of California Press, Berkeley, 2007.

[37] David Arnold, *Famine: Social Crisis and Historical Change*, Basil Blackwell, Oxford, 1988; R.J. Fisher, *If Rain Doesn't Come: An Anthropological Study of*

investigation is gaining legitimacy. Unexplored dimensions of the physical world are being continuously investigated and integrated with the historical explanations. Beginning with the history of agriculture and agrarian history, historians have gradually started acknowledging the significance of other environmental factors such as ocean, fertility of the soils, forest, water, etc.[38]

ENVIRONMENT AS A FACTOR IN WRITINGS ON THE INDIAN SUBCONTINENT

Till recently there was almost a total absence of academic work on the nature and significance of environmental concerns in general and climatic factor in particular in India.[39] Mostly, the historians working on the history of India have ignored the ecological dimensions of social life. The prime concerns, more or less exclusively, have been on relations of production and distribution of production and never on the ecological fabric within which these are embedded, and which these in turn transform.[40] The recent writings on environmental concerns in India have tended to concentrate on the exploitative nature of British colonialism. The colonial policies were termed as the major cause of environmental degradation, especially forest wealth. Thus it is not surprising that most of the writings on environment in India choose 'colonialism' as the starting point and inevitably centre on it and by implication on industrial development, as causing a serious disruption in the

---

Drought and Human Ecology in Western Rajasthan, Manohar, Delhi, 1997; A.R. Vasavi, *Harbingers of Rain: Land & Life in South India*, Oxford University Press, Delhi, 1999.

[38] Fernand Braudel, *The Mediterranean and the Mediterranean World in the Age of Philip-II*, vol. I, tr. Sian Reynolds, Collins, London, 1981; vol. II, tr. Sian Reynolds, Collins, London, 1982; Marc Bloch, *Feudal Society*, tr. L.A. Manyon, Routledge and Kegan Paul, London, 1961; Richard Grove, *Green Imperialism: Colonial Expansion, Tropical Island Edens and the Origins of Environmentalism, 1600–1860*, Cambridge University Press, Cambridge, 1995.

[39] A good survey of the emergence and growth of environmental history can be seen in the following articles: Grove and Damodaran, 'Imperialism, Intellectual Networks', op. cit.

[40] Guha and Gadgil, *This Fissured Land*, p. 6.

otherwise cordial and mutually advantageous relationship between environment and society. In such a discourse the traditional practices have been seen as environment-friendly just because the 'modern' practices have been termed as exploitative.[41] This tendency, thereby placed the primary agency with 'colonialism' and inadvertently undermined the significance of the physical world. Further, offering the agency to 'colonialism' has indirectly put the examination of the 'human ecological footprint' at the centre-stage.[42] A deeper integration of variations in the climatic/environmental factor in influencing the historical processes has been a casualty. Wider interaction with climatologists working on the history of the climate of India has often been relegated to

[41] Most of the works establish a dichotomic relationship between traditional practices and 'modern' practices. There is not enough suggestion that the attributes so ascribed to the traditional practices are in fact based on a serious, analytical explanation of contemporary evidence. The non-exploitative character of traditional social practices *vis-à-vis* the environment becomes an axiom.

[42] The domination of these issues in environmental history can be gauged by the following acceptance: 'We make no apology for devoting so much of the book to the history of the relationship between forests, people and the state, and to the history of the discourse and ideology of colonial forestry in India, Burma and Malaysia. At the peak of its power the Indian Forest Department, for example, directly controlled over one-fifth of the land area of South Asia. Moreover, the forest history of the subcontinent and South Asia varies enormously from area to area, and we feel it necessary to highlight these differences and make a start at producing a series of detailed and empirical environmental histories, concentrating quite deliberately on the forest sector.' Richard H. Grove, Vinita Damodaran and Satpal Sangwan, 'Introduction' in Richard H. Grove, Vinita Damodaran and Satpal Sangwan (eds.), *Nature and The Orient*, Oxford University Press, Delhi, 1998, p. 2. It was aptly challenged by Neeladri Bhattacharya; 'if the old agrarian history neglected the forests and pastures, environmental history now has banished the peasants fields and farms from the realm of historical concerns. It was as if these histories were not connected, as if environment did not impinge on the agrarian world.' Neeladri Bhattacharya, Introduction, *Studies in History*, vol. XIV, no. 2, n.s. 1998, pp. 1–2. The two gaps, the neglect of pre-colonial historical themes and the intimate connections of fields and forests, were not unrelated. Furthermore, another important collection of essays also explains changes during and after the colonial period in India in terms of the 'human ecological footprint': Mahesh Rangarajan, and K. Sivaramakrishnan (eds.), *India's Environmental History*, vols. I and II, Permanent Black, Ranikhet, 2012.

the field of policy makers.[43] Recently, while offering a preliminary survey of 'Natural Disasters', Tirthankar Roy has argued that there is dearth of writings on the role, significance and implications of natural disasters on the society in general and social responses in particular.[44]

The situation is no better with respect to writings on pre-colonial Indian history with respect to according space to environmental concerns in general and climatic concerns in particular.[45] Although one can discern that in the past few decades there has been a gradual shift in the historians' concerns from mere political history to economic and social history, nevertheless, these shifts have taken place without giving even peripheral attention to their ecological context. Unfortunately, issues and concerns like an interface between environment and contemporary social patterns or the relationship between environment and socio-economic practices and processes of negotiations and adaptations have received only cursory mention.

Nevertheless, a broad survey of the writings on the environmental concerns in India will help us locate the relevance of the present work. Pre-colonial writings on environment have primarily addressed the issues pertaining to forest, flora-fauna, forest dwellers, pastoral communities, etc. Francis Zimmermann has examined ancient texts to construct the ecology of the period. He has questioned the practice of equating the term *jungle* with forest. Zimmerman has explored the suggestive ecological references from the ancient texts where wild landscapes are classified in two groups: *jungal*, 'those of the dry lands', and *anupa*, 'those of the

[43] Deep Narayan Pandey, Anil K. Gupta and David M. Anderson, 'Rainwater Harvesting as an Adaptation to Climate Change', *Current Science*, vol. LXXXV, no. 1, 10 July 2003, pp. 46–59. Climatologists working on past climate of India are primarily trying to figure out the future projections and thereby examining in terms of future planning.

[44] Tirthankar Roy, *Natural Disasters and Indian History*, Oxford University Press, Delhi, 2012.

[45] Even a cursory look at the collections carrying articles on the environmental history of India makes it amply clear that the medieval period has been a major casualty in the field of environmental history. See Grove et al., *Nature and the Orient*; Rangarajan and Sivaramakrishnan, *India's Environmental History*.

marshy lands' and pointed out that by closely examining such texts we can infer a great deal about ancient ecology.[46] Following a more conventional path, Aloka Parasher-Sen has studied the Mauryan period to 'understand how the state perceived the forest dwellers and sought to subordinate and assimilate them. Geography and the perceived existence of the hostile tribes defined the frontiers of the empire and both had to be mastered for the expansion and integration of the state.'[47] The other major concern has been the study of social formations and it has been influenced by the methodologies and tools deployed by anthropology and archaeology.[48] Ranabir Chakravarti has highlighted the role of hydraulic management in the process of settlement in the ancient period.[49]

As we move to medieval India we discover a general dearth of scholars focusing on environment and on human–environment interaction. We may refer to the two initial chapters in *The Cambridge Economic History of India*, vol. I[50] by Irfan Habib and Burton Stein on 'The Geographical Background' especially of north India and 'South India: Some General Consideration of the Region and its Early History', respectively as studies located on the margins of environmental history. In fact, closest to the field of environmental history can be seen in the writings which emerged

[46] Francis Zimmermann, *The Jungle & the Aroma of Meats: Ecological Themes in Hindu Medicine*, University of California Press, Berkeley, Los Angeles, London, 1987. Similar trends are visible in Roger Jeffery (ed.), *The Social Construction of Indian Forests*, Manohar, Delhi, 1998.

[47] Aloka Parasher-Sen, 'Of Tribes, Hunters and Barbarians: Forest Dwellers in the Mauryan Period', *Studies in History*, vol. XIV, no. 2, n.s. 1998, pp. 173–91. Also Shereen Ratnagar, 'Pastoralism as an Issue in Historical Research', *Studies in History*, vol. VII, no. 2, n.s. 1991, pp. 181–93.

[48] R. Ray, *Ancient Settlement Patterns in Eastern India: Prehistoric Period*, Pearl Publishers, Delhi, 1987; M.L.K. Murthy, 'Environment, Royal Policy and Social Formations in the Eastern Ghats, South India', *Proceedings of the Indian History Congress*, Delhi, 1993, pp. 615–31; D.K. Bhattacharya, *Ecology and Social Formation in Ancient History*, K.P Bagchi & Company, Delhi, 1990.

[49] Chakravarti, 'The Creation and Expansion'; 'Agricultural Technology in Early Medieval India.'

[50] Tapan Ray Chaudhuri and Irfan Habib (eds.), *The Cambridge Economic History of India*, vol. I, Cambridge University Press, Cambridge, 1982.

broadly under the influence of the *Annales* school of historical writings. Harbans Mukhia's *Was There Feudalism in Indian History?*,[51] appears as a major departure from earlier writings where the ecological context—soil productivity, and climatic setting—was accorded importance in historical explanations. Further, impetus to the significance of geographical factors in historical investigation, especially in the medieval period, was offered by translations of the writings of *Annales* by Aymard and Mukhia.[52]

Further developments in the field can be located in terms of regional histories;[53] Chetan Singh carrying the legacy of Mukhia integrated the significance of the ecological setting in his investigation of historical developments in seventeenth-century Punjab.[54] Chetan Singh has explored the relationship between environment and society in the western Himalayas. He argued that:

> . . . my project rested on the belief that there were some long established and well understood relationships between society and its physical surroundings. . . . Such fundamental relationships did indeed exist: a society could hardly have survived for any length of time without them. It was, however the clear-cut enunciation of these relationships that was missing. This required the deliberate elaboration both of socio-economic processes and the specific ecological environment within which they operated.[55]

The significance of the ecological context in the historical investigation of medieval India received a tremendous boost by the works of Richard Eaton.[56] He suggested that medieval ecologies

[51] Mukhia, *Was There Feudalism in Indian History*, op. cit.

[52] Maurice Aymard and Harbans Mukhia (eds.), *French Studies in History*, vol. I: *The Inheritance*, Orient Longman, Delhi, 1988 and vol. II: *The Departure*, Orient Longman, Delhi, 1990.

[53] Guha and Gadgil have argued that ultimately the ecological history of India must be constructed around detailed regional studies. Guha and Gadgil, *This Fissured Land*, p. 6.

[54] Chetan Singh, *Region and Empire: Punjab in the Seventeenth Century*, Oxford University Press, Delhi, 1991.

[55] Chetan Singh, *Natural Premises: Ecology and Peasant Life in the Western Himalaya 1800–1950*, Oxford University Press, Delhi, 1998.

[56] Richard Eaton, *The Rise of Islam and the Bengal Frontier, 1204–1760*, Oxford University Press, Delhi, 1997. On the fringes one can locate the followings

were not static and human adaptations to the changes need to be accorded due recognition in historical writings. Further, he aptly argued that even the state in medieval times was aware of the significance of the ecological context and often offered assistance in mitigations, if necessary.

Another trajectory within the wider writings on the environmental history of pre-colonial India has been offered by Sumit Guha. He has very articulately argued that the historical investigation of the pre-colonial past in the Indian subcontinent should not borrow categories from the works dealing with colonial and post-colonial periods in Indian history. It is unfortunate to note that the seminal work on the ecological history of India by Madhav Gadgil and Ramchandra Guha as mentioned earlier had depicted the pre-colonial past as an era of 'prudent resource-use' in contrast to the exploitation of natural resources during colonial practices. Challenging such generalizations, Sumit Guha pointed out:

> It must be evident that Gadgil's model breaks down at every point; small endogamous groups did not occupy the same location for long periods of time, and even when they did, they had lacked exclusive access to particular resources. The human species is too opportunistic and too versatile to be confined to particular niches for centuries on end.[57]

Further, he went on to examine the problematic understanding of different categories designating relationships between particular sections of the society with the specific ecological setting in *Environment and Ethnicity*. He argued that one needs to revisit the categorization of different communities which has unfortunately been done merely on the basis of resource-use practices often ignoring the complex connection with the wider social and ecological settings. He also pointed out that modern categorization primarily executed under the British colonial rule has often reduced their identity in terms of resource-use practices

---

as well: Shireen Moosvi, 'Ecology, Population Distribution and Settlement Pattern in Mughal India', *Man and Environment*, vol. XIV, no. 1, 1989, pp. 109–16; Mohd. Afzal Khan, 'Environment and Pollution in Mughal India, *Islamic Culture*, vol. LXXVI, no. 1, January 2002, pp 101-16.

[57] Guha, *Environment and Ethnicity in India*, p. 43.

and thereby negated possibilities of a wider social role to these communities.

Given the larger texture of concerns on the pre-colonial past it is not surprising to note that only a few writers have probed the significance of pre-colonial water systems,[58] this is particularly true of north and north-west India.[59] Kathleen Morrison as part of a larger project to examine the water systems in and around Vijayanagara during the sixteenth and seventeenth centuries has documented extension of agriculture in the marginal areas. Monsoon rains were being increasingly captured to develop the agrarian potential of even the areas on the fringes of agrarian landscape. Most of the times, it was the paucity of water that had restricted the agrarian production and with the channelization of monsoon-rains expansion of agrarian settlements became stable.[60] On the other hand, Devra has argued that one can decipher a gradual extension of the desert in northern Rajasthan towards Punjab. He has suggested that a gradual shift in the river courses led to changes in the climate-landscape of the region.[61] At times

[58] David Ludden, 'Ecological Zones and the Cultural Economy of Irrigation in Southern Tamilnadu', *South Asia*, vol. I, no. 1, 1978, pp. 1–13; Burton Stein, *The New Cambridge History of India*, vol. 2: *Vijayanagara*, Cambridge University Press, Cambridge, 1994; *Peasant State and Society in the Medieval South India*, Oxford University Press, Delhi, 1980.

[59] Tripta Wahi, 'Water Resources and Agricultural Landscape: Pre-colonial Punjab', in Indu Banga (ed.), *Five Punjabi Centuries: Polity, Economy, Society and Culture, c.1500-1990*, Manohar, Delhi, 1997.

[60] Kathleen D. Morrison, 'Production and Landscape in the Vijayanagara Metropolitan Region: Contributions of the Vijayanagara Metropolitan Survey', in J.M. Fritz, T. Raczek, and R. Brubaker (eds.), *Vijayanagara: Archaeological Exploration 1990-2000: Papers in Memory of Channabasappa S. Patil*, Vijayanagara Research Project Monographs, vol. X, Manohar and AIIS, 2005, pp. 423–34; *Fields of Victory: Vijayanagara and the Course of Intensification*, Munshiram Manoharlal, Delhi, 2000.

[61] G.S.L. Devra, 'Problems in the Delimitation of the Rajputana Desert During the Medieval Period', in Rakesh Hooja and Rajendra Joshi (eds.), *Desert, Drought & Development: Studies in Resource Management and Sustainability*, Rawat Publications, Jaipur, 1999, pp. 371–82; 'Physiography and Environment of Thar Desert of Indian Sub-continent During Seventh and Eighth Century', Draft paper presented at the 6th International Conference on 'Perspective on Rajasthan'

he challenges the findings of Dhavalikar who has argued that the Indian subcontinent witnessed climatic variability between the seventh and eleventh century. Along with shifts in the river courses, Dhavalikar suggests that perhaps this era witnessed a change in the climatic patterns.[62] In an important contribution, Meena Bhargava has examined the processes of social negotiations to the changes in the river courses.[63]

Traditional water harvesting systems have been another area of investigation. The role of state in augmenting systems of water-harvesting for agrarian production has been examined by H.C. Verma for thirteenth to sixteenth centuries.[64] On the basis of extensive exploration of epighraphic, archival and literary sources, water Harvesting and conservation practices in Mewar region has been very eloquently documented by B.L. Bhadani.[65] On a different trajectory, for the later period, Yogesh Sharma has documented the perceptions of European travellers with respect to artificial water bodies in India.[66] Day-to-day social negotiations with the environment have been explored by me in couple of articles to argue against the notion that the traditional societies always lived in harmony with nature. I have cited several cases of exploitation of nature by traditional societies in Rajasthan.[67] Further, I have

---

organized by Institute of Rajasthan Studies, Jaipur, Rajasthan, India (2–3 January 2010).

[62] Dhavalikar, 'The Golden Age'; 'Green Imperialism'.

[63] Meena Bhargava, 'Changing River Courses in North India: Calamities, Bounties, Strategies Sixteenth to Early Nineteenth Centuries', *The Medieval History Journal*, vol. X, nos. 1 & 2, 2007, pp. 183–208.

[64] H.C. Verma, *Harvesting Water and Rationalization of Agriculture in North Medieval India: Thirteenth-Sixteenth Centuries*, Anamika, Delhi, 2001

[65] B.L. Bhadani, *Water Harvesting, Conservation and Irrigation in Mewar (AD 800-1700)*, Manohar, Delhi, 2012.

[66] Yogesh Sharma, 'The Circuit of Life: Water and Water Reservoirs in Pre-modern India', *Studies in History*, vol. XXV, no. 1, n.s. January-June 2009, pp. 69–108.

[67] Mayank Kumar, 'Claims on Natural Resources: Exploring the Role of Political Power in Pre-Colonial Rajasthan, India', *Conservation and Society*, vol. III, no. 1, June 2005, pp. 134–49. However, one must be careful that the magnitude of exploitation of natural resources did multiply manifold under the impact of the Industrial Revolution and have gradually attained devastating proportions.

also argued that it is very necessary to accord primacy to the water resources while examining the relationship between environmental features and the settlement patterns.[68]

The changing history of the encounters of humans and animals has become another field of growing interest, both in terms of changing elite taste and of ground level conflicts and co-existence. In this context, *The End of Trail: The Cheetah in India*[69] stands out. The author has attempted to trace the history of the cheetah in India, its origin, spatial distribution, attitude towards the animal, gradual erosion of space for the big animal and finally the extinction of the species. It is an important contribution that helps in a comprehension of complex relationships between fauna and the society, especially the explanation that the extinction of the animal was caused by the side-effect of the larger historical process and not as a direct process of elimination of the species as it was for other 'big game'. Another monograph by Divyabhanusinh on lions carries the quest further to explore the fauna in their own world.[70] In a very recent article Pratyay Nath has argued that Mughal warfare practices evolved in response to a very different landscape of Ganga–Yamuna Doab.[71] Negotiating with the landscape where river and water-channels criss-cross the region, Mughal gradually started relying more on the elephants in the battles. Thus, he brings out a very different set of social relationships with fauna and its

[68] Mayank Kumar, 'Situating the Environment: Settlement, Irrigation, and Agriculture in Pre-colonial Rajasthan', *Studies in History*, vol. XXIV, no. 2, n.s., 2008, pp. 211–33; 'Flexibility and Adaptability: Agrarian Expansion and Traditions of Water Management', *The ICFAI University Journal of History and Culture*, III, no. 3, 2009, pp. 31–55; 'Ecology and Traditional Systems of Water Management: Revisiting Medieval Rajasthan', in Mahesh Rangarajan (ed.), *Environmental Issues in India*, Pearson Longman, Delhi, 2006, pp. 70–96.

[69] Divyabhanusinh, *The End of a Trail: The Cheetah in India*, Oxford University Press, New Delhi, 1999.

[70] Divyabhanusinh, *The Story of Asia's Lion*, Marg Publications, Mumbai, 2005.

71 Pratyaya Nath, 'Rethinking Early Mughal Warfare: Babur's Pitched Battles, 1499-1529', in Raziuddin Aquil and Kaushik Roy (eds.), *Warfare, Religion and Society in Indian History*, Manohar, Delhi, 2012, pp. 109–45.

significance.[72] Other interesting genre of writings has examined the Princely States' interactions with colonial power as mediated by the flora-fauna of their respective regions. Barbara Ramusack has explored the British hunting practices and its appropriation and reinterpretations by Indian Princes.[73]

Last but not the least, it is necessary to acknowledge the historians working on the agrarian history of medieval India, especially for Rajasthan. The writings of W.H. Moreland set the tone of agrarian studies for medieval India[74] while *The Agrarian System of Mughal India 1556–1707* remains the most authoritative work on the medieval agrarian structure and agrarian relations.[75] Similarly, for Rajasthan one cannot ignore the contributions of G.D. Sharma,[76] S.P. Gupta,[77] G.S.L. Devra,[78] Dilbagh Singh,[79] B.L. Bhadani,[80] etc. These works have offered a comprehensive profile

[72] Pratyay Nath, 'From Mocking to Becoming the 'Lords of the Elephants', War-Elephants and the Mughals, 1526–c. 1590', paper presented at a Workshop on Environments and Histories, organised by Nehru Memorial Museum & Library, New Delhi, 12 May 2012.

[73] Barbara N. Ramusack, *The Indian Princes and their States*, 1st Asian edition, Cambridge University Press, New York, 2005; Anand S. Pandian, 'Predatory Care: The Imperial Hunt in the Mughal and British India', *Journal of Historical Sociology*, vol. XIV, no. 1, March 2001, pp. 79-107; John M. MacKenzie, *The Empire of Nature: Hunting, Conservation and British Imperialism*, Manchester University Press, New York, 1988.

[74] W.H. Moreland, *India at the Death of Akbar: An Economic Study*, Reprints and Trans Publications, Delhi, 1974; *Agrarian System of Moslem India: A Historical Essay with Appendices*, Low Price Publications, Delhi, 2003.

[75] Irfan Habib, *The Agrarian System of Mughal India 1556–1707*, 2nd rev. edn., Oxford University Press, Delhi, 2002.

[76] G.D. Sharma, *Rajput Polity: A Study of Politics and Administration of the State of Marwar, 1638–1749*, Manohar, Delhi, 1977.

[77] S.P. Gupta, *The Agrarian System of Eastern Rajasthan (c. 1650–1750)*, Manohar, Delhi, 1986.

[78] G.S.L. Devra, *Rajasthan ki Prashashnik Vyavastha*, Dharti Prakashan, Bikaner, 1981.

[79] Dilbagh Singh, *The State, Landlords and Peasants*, Manohar, Delhi, 1990; 'The Role of Mahajans in the Rural Economy in Eastern Rajasthan During the 18th Century', *Social Scientist*, vol. II, no. 10, October 1974.

[80] B.L. Bhadani, *Peasants, Artisans and Entrepreneurs: Economy of Marwar in the 17th Century*, Rawat Publication, Jaipur, 1999.

of agrarian production in the given landscape of their area of research along with enunciation of socio-political interventions deployed by the society. While much of this work has been useful and important, there are some significant gaps in research for this region. An inherent fallacy has been the false segregation between the concerns of the ruling elites and the rural society. It is assumed that actual negotiations with the ecological niche were the concerns of rural society, which was negotiating with the vagaries of nature, and political powers were not at all concerned with it. The knowledge of ecological cycles, tools of production, and processes of agrarian production were treated as the exclusive domain of rural society and political formations were simply interested in revenue appropriation. In the recent past, a few scholars with anthropological orientations have explored this neglected area of historical significance.[81] None the less, there is a need to revisit the historical process of human adaptations to the climatic variability in the light of recent writings drawing attention towards the importance of the ecological context.

A NOTE ON THE EVIDENCE

It will not be inappropriate to confess that the greatest hindrance while attempting this work has been the limitation of conventional sources of history, often forcing me to transcend either the disciplinary boundaries or temporal and geographical settings. Nevertheless, let me put on record that I have been benefitted by

[81] R.T. Rosin, *Land Reforms and Agrarian Change: Study of a Marwar Village from Raj to Swaraj*, Rawat Publication, Jaipur, 1987; Fisher, *If Rain Doesn't Come*. Other anthropological studies in Rajasthan are those by G.R. Gupta, *Marriage, Religion and Society: Patterns of Change in an Indian Village*, Sterling Publishers, Delhi, 1976; P.C. Aggarwal, *Caste, Religion and Power: An Indian Case Study*, Shri Ram Centre for Industrial Relations, Delhi, 1971; A. Chakravarti, *Contradiction and Change: Emerging Patterns of Authority in a Rajasthan Village*, Oxford University Press, Delhi, 1975; N.S. Jodha, *Life on the Edge: Sustaining Agriculture and Community Resources in Fragile Environments*, Oxford University Press, Delhi, 2001; A.G. Gold and B.R. Gujjar, *In The Times of Trees and Sorrows: Nature, Power and Memory in Rajasthan*, Duke University Press, Durham & London, 2002.

both. Given the visible invisibility of climate and landscape in most of the records of the past it was contingent upon me to substantiate written sources with archaeological evidences. The disciplinary boundaries were fruitfully transcended to gather records of the past climate. My nine-month stint as Fulbright Fellow at the Decision Centre for Desert City exposed me to concerns and tools deployed by the climatologists and in what ways their conclusions were integrated by social scientists. Studies of the past climate usually take shape on a longer time-scale and on a bigger spatial domain and at times may not respond to your chosen area of research. Thus, one needs to often rely on comparative analysis which is to be corroborated against the 'hard' historical evidences from one's area of research for better depiction of climate of the region.

The sources can be clubbed under two broad categories—literary and archaeological—with epigraphic sources staking claim to both the categories. 'From a historian's point of view, literary sources include texts—long or short, written or oral; archaeological sources include all tangible, material remains. But these distinctions are not absolute.'[82] Due to limitation of the sources the present work relies more on the secondary works for the early medieval times. It is important to note that these works for obvious reasons have relied heavily on epigraphic sources. Epigraphic sources 'are usually contemporaneous to the events they speak of and their information can be connected to a time and place . . . inscriptions often reflect what people were actually doing.'[83] Furthermore, most of the evidences examined in the secondary works usually offer indirect evidences for the climate of the era. These sources have been corroborated against the literary accounts for the period, which comprise another set of documents.

None the less, evidences of climatic conditions can be culled by reorienting our examination of the literary and archival sources as well. For the later period, the present work has extensively relied on the primary sources. Archival and literary sources

[82] Upinder Singh, *A History of Ancient and Early Medieval India: From the Stone Age to the 12th Century*, Pearson Longman, Delhi, 2008, p. 13.

[83] Ibid., p. 49.

offer immense insight to realize the centrality of monsoon in the popular perceptions and social concerns. There are a variety of records available at the Rajasthan State Archives in Bikaner. It is a rich repository of official and non-official records pertaining to various principalities of pre-Independence Rajasthan. The archives are especially rich in records in official languages of different regional states. It has got a good collection of Persian records also, primarily Mughal *firmans* and royal correspondence with the Mughal emperors. The archival records are often in the nature of official correspondence between the ruler on the one hand and the officials posted at different parts of the concerned state on the other. Some of these were part of the royal communication system and some were official directives.

An important category of official documents available in archival repositories is the *arzdashts* (petitions) written by the *amils*, *fauzdar* or other *pargana* officials to the ruler at Amber. These officials regularly reported to the ruling authorities, about various revenue and administrative details of the areas under their control. They were written in response to the problems faced by peasantry and officials and therefore provide useful insights into the various concerns of the state and the common man. Simultaneously, these also give us the responses/remedies evolved to solve the problem. The *arzdashts* have been a very important source for us, as they bring to the fore the problems faced at the most basic unit of production, that is, the village. The only limitation of these sources has been the lack of any sequence and non-availability of corresponding imperial response to the queries of the officials.

The other important archival source used in the study are the *arhsattas*. These are basically ledgers of receipts and expenditure maintained at the *pargana* level. The *arhsattas* are known to have contained basically 68 (*arhsat*) categories of information and were thus called *arhsattas*. We get detailed information on the crops grown in a *pargana* and this information is available for both cropping seasons. The measurements of cultivable land, that is, land under actual cultivation, the tax thereupon and other taxes realized in a *pargana* as was deemed necessary by the state are also provided in these documents. Of the 68 categories of information, one is known as *hasil-farohi*, which means the collection of money, deposited as fine in the court at the *pargana* level. The documents of this category also provide us information on the various crimes committed in the region. Interestingly most of the crimes were committed against the common environmental property resources, that is, felling of green trees, etc.

In the archival collections we also get an arrangement of loose papers put together chronologically. These are known as *chitthis*. These are letters written by the *diwan* of Amber mainly to correspond with the *amils* and other officials of the *pargana*. The substance of the specific information and complaints received by the *diwan* as contained in the *chitthis* are also followed by instructions on the matters raised in the *chitthis*. This gives us a reading of the nature of hardships faced by the peasants and the attitude of the state towards different social groups in disposing the grievances. At the time of drought, the migration of peasantry was a normal practice and the state extended various concessions to retain them. These documents contain very important information regarding the state's response to various environmental distresses.

Vakil reports are the second set of important archival documents. These are, in fact, the correspondence between the ruler's representative posted at the Mughal court and the ruler at Amber. The *vakil* reports also kept the ruler informed about the political developments at the Mughal court. These also record the complaints made to the Mughal emperor of atrocities on peasants. The same documents also record the fact that these matters were taken up by the Mughal emperors with the respective *vakils*. This reflects the concern of the central authorities about the conditions of the peasantry in Rajasthan.

In addition to these we also get *taqsim* records providing a summary of revenue details maintained at the *pargana* level. These documents provide numerous other details too, like the total area under cultivation, area of cultivable/uncultivable land, area under habitation, forest, drain and rivers, etc. The evidence of this kind is extremely useful in sketching the environmental resources of the region at some length. Not many sources give these informations in such a direct manner, as do the *taqsim* records. In fact, *taqsim* records also reveal the nature of relationship that existed between political exigencies of the state and their adjustment/ accommodation with the resources made available by the nature there.

We also have a kind of memorandum or a document of remembrance called the *yaddashtis*. These were written at the *pargana* level by the revenue officials like the *chaudhuri*, *qanungo*, *patel* and *patwari*. *Yaddashtis* provide details related with the settlement of the village, that is, when and who settled it, etc. This category of document gives us a historical account of the villages and the process of settlement. We are thus enabled to see a picture of the ecological history of the place, even if it is a little hazy. Not many archival documents are so dense when it comes to the historical tracting of the region and its environmental resources. *Yaddashtis* also contain detailed information about arable and fallow land and the livestock and the bullocks owned by the peasants of the village or the *pargana*. Also listed are the details of the availability of agricultural tools with the peasants.

As far as the Marwar region is concerned, we have used a very informative document known as the *sanad parwana bahi*. The term *sanad* and *parwana* explain the functions this document performed. These are primarily imperial directives issued to the *pargana* officials in response to various complaints and representations received by the state. The subject matter of these documents covers a very large domain. It ranges from routine complaints of undue collection of revenue against the state officials to the mutual disputes related even to the amount of water of a well to be shared among them. As the environment had imposed limitations on resource generation in the region, there were conflicts over distribution of these resources. These limitations compelled the state to extend several concessions in times of natural distress. These documents are very important as they also highlight the various concessions offered by the state in times of distress, the incentives to be offered for rehabilitation of the deserted village to check the migration due to famine or drought, highlighting the concerns of the state in this desert region.

The *kagad bahi* of Bikaner is particularly significant for a better comprehension of the land revenue system. A good deal of light is thrown on the problems faced by the peasants, the primary producing class, and the methods of redressal adopted by the state. Various kinds of concessions and help were offered by the state to the peasantry, as monsoon failures were a frequent phenomenon in this region. The details of various concessions offered to the traders have also been recorded in the *kagad bahi*.

Marwar Pargana ri Vigat, written by Diwan Munhot Nainsi, *diwan* of Marwar state, is an invaluable resource for this study. It is basically an administrative manual providing a detailed description of the territory, soils, nature of habitation, agricultural types, nature and number of wells, other sources of water, caste and class composition of population, etc. It is such an informative document that details of villages, which were deserted, and later rehabilitated, also find mention here.

For Jaisalmer we have a similar, though from late nineteenth-century period, source called *Tawarikh Jaisalmer*. It is an official history of the Jaisalmer State written by Diwan Nath Mal in 1879. *Tawarikh* was an attempt to provide an administrative manual on the lines of *Marwar Pargana ri Vigat*, and contains details of the wells in villages, the methods of irrigation, details of crops grown in the region, etc. It is of immense importance for the present study as the single-most valuable source for the region of Jaisalmer.

Another late nineteenth-century official history—*Vir-Vinod*—has been used for the Mewar region. It is somewhat similar to *Tawarikh Jaisalmer* in design, but more comprehensive in detail, as it provides the history of various regional states of Rajasthan on the lines of Nainsi's *Khyat*. Here was an attempt to provide an administrative manual on the lines of *Marwar Pargana ri Vigat* for the Mewar region. The details of agricultural production, construction of irrigational dams, etc., have been described in the text. The geographical features of Mewar have also been elaborately described here. There are different ways in which historians have used different kinds of sources to construct the history of the pre-colonial past. Here an attempt is made to revisit the sources once again in the light of fresh research and findings to bring out the centrality of the monsoons.

PLAN OF THE BOOK

It will not be an exaggeration to state that even today the monsoon characterizes Indian climate. Prognostication of monsoons has dominated the concerns of modern climatologists and predictions have been eagerly awaited along with farmers, by the policy makers since monsoon forms the backbone of Indian agriculture. Any disruption in the nature, timing or spatial distribution of rainfall across the subcontinent is seen like a catastrophe. Developments in the science of climatology have made tremendous progress in the prediction of changes/variance in temperature on a long-term basis but have not been very successful in the prediction of changes/variation expected in precipitation.[84] Moreover, there has been a general scarcity in terms of historical investigations of the climate patterns, especially for the medieval centuries. Thus, the present book is divided into two broad sections; 1. Monsoon Landscape and 2. Rain Catchers.

Chapter 1 offers a sketch of the regional landscape in relation to the climatic characteristics. In popular perceptions, Rajasthan emerges as a singular entity, often obscuring the physical and cultural diversity of Rajasthan. Such a characterization has also obscured the diversity in human–nature interactions in different parts of Rajasthan. The monsoon, especially in Rajasthan, is confined to four months only, with some rains in winter due to western disturbances over the Arabian Peninsula. The western side—Jaisalmer, Bikaner and a greater part of Marwar—comes very near the limits of that part of Asia that belongs to the rainless district of the world, yet the south-western monsoon winds bring annually a little rain from the Indian Ocean which scarcely averages more than 5 inches per annum. In this chapter an attempt has

[84] Andrew W. Ellis, et al., 'Estimating Future Runoff Levels for a Semi-arid Fluvial System in Central Arizona, USA', *Climate Research*, vol. XXXV, 2008, pp. 227–39; K.S. Kavi Kumar, 'Climate Sensitivity of Indian Agriculture', Working Paper 43/2009, Madras School of Economics, Chennai, 2009.

been made to decipher the relationship between the physical and biological settings in the larger context of monsoon ecologies and the emergence of regional identity, if any. In what ways regional polities have drawn on the climate and landscape to define their identity has been another concern of the chapter. Moreover, despite diverse landscapes in different parts of Rajasthan, in what ways monsoon has provided a unifying thread to substantiate the political consolidation is also hinted at.

The monsoon is the backbone of civilization in India. The term 'monsoon' perhaps has its origin in the Arabic term 'mausim'. Monsoon is basically any type of major wind system that seasonally reverses its direction. In India it is erroneously presumed that the monsoon is an exclusively Indian phenomenon. Reversals of major wind-systems like monsoons are also apparent along the Gulf coast of the USA and in Central Europe as well as in Africa and Eastern Asia. What causes the monsoon or reversal of wind systems is a difference of annual temperature trends over land and sea along with numerous other factors. In ancient times, monsoon was explained in terms of the then comprehensible factors like stars, constellations, the moon, the sun, etc. It is suggested that Indian cosmology was excessively obsessed with the celestial bodies and its influences on human existence. The sun, moon and seasons showed a remarkable cyclical pattern. Similar cyclical patterns were also visible in the wider space, so it was but natural for them to explain changes in climate only in terms of noticeable objects/features. The second chapter also points out the centrality of the monsoon in popular sayings in terms of rituals and day-to-day affairs. Moreover, this chapter also argues that in the absence of tools to measure numerous features of the climate, attempts were made to explain climatic features in very general and vague terms. The uncertainties associated with the monsoon were usually explained as the influence of constellations or heavenly bodies. Even minor temporal or spatial variations have played havoc with the lives of millions. Historically, it has been demonstrated that any temporal or spatial disturbance in the monsoon-induced rainfall have gravely affected agrarian production in different regions and thereby have caused famines.

According to Sanskrit literature as well as to folk knowledge, the monsoon is indeed the result of a periodic reversal, but the realization of this reversal is with the sun and rains, not the winds. In the view of priests, astrologers and physicians—who were responsible for establishing the cycle of the festivals, the calendar of agricultural activities, the seasonal diets, and regimens—there exists a fundamental alteration between the Dry and the Wet principles. Monsoon comes after the summer solstice, when Agni (the sun) reverses its course and the earth comes under the rule of Soma (the moon), the dispenser of rain.[85]

The second section of the book historicizes the monsoon in terms of historical processes. Like any other region solely dependent upon the monsoon, the climatic variability in the case of Rajasthan also is greatly associated with the timely onset and regular spread of the monsoon. There is not a single perennial river in the north and north-west region and ground water is available at great depth and is often brackish.[86] It has been suggested that this was not the case earlier when at least two rivers flowed through this region and sustained human settlements.[87] It has been a matter of debate to argue the relative significance of these rivers with respect to their contribution to the growth and/or decline of the urban settlements or who were the architects of these settlements.[88] To me what is more relevant is the fact that general dry climatic conditions in the region were aggravated by a gradual shift in the course of these two rivers perhaps due to tectonic movements.[89] However, the dry climatic conditions in the region during early medieval period cannot be ignored and demand an investigation in the process

[85] F. Zimmerman, 'Monsoon in Traditional Culture', in Jay S. Fein and Pamela L. Stephens (eds.), *Monsoon*, John Wiley & Sons, New York, Chichester, Brisbane, Toronto, Singapore, 1987, p. 51.

[86] Munhot Nainsi, *Munhot Nainsi ri Khyat*, vol. II, Badri Prasad Sakaria, Rajasthan Oriental Research Institute, Jodhpur, 1962, p. 286.

[87] A. Ghosh, 'The Rajputana Desert: Its Archaeological Aspects', in S.P. Gupta (ed.), *A Stein: An Archaeological Tour Along the Ghaggar-Hakra River*, Kusumanjali Prakashan, Meerut, 1989, pp. 98–106.

[88] T.K. Rajalakshmi, 'The Riddle of a River', *Frontline*, vol. XIX, no. 16, 3–16 August 2002. Retrieved on 9 March 2010 from http://www.thehindu.com/fline/fl1916/19160970.htm.

[89] Ghosh, 'The Rajputana Desert'.

of adaptation which resulted in the growth of settlements with sufficient agrarian base. Gradual non-availability of a perennial source of water is reflected in a new set of negotiations between this seasonal availability of water and the settlement patterns. This adaptation which also took care of the flux associated with water which is usually more pronounced in monsoon-dependent economies is the focus of Chapter 3.

Chapter 4 studies the seasonal availability of water in monsoon ecologies. Seasonal availability induced human ingenuity to evolve numerous indigenous methods of water conservation to ensure availability for the whole year. Limited availability of water only during the four months of monsoon necessitated water conservation. The conservation of water was necessary not only for potable purposes but also for agriculture. Moreover, in Rajasthan, the water table varied a lot and was brackish in the western part for most of the times. As one moves away from the Aravallis in the broad belt across the desert, the water table sinks rapidly. The depth of water in the west and north-west varies from 50 to 115 m. Apart from more traditional and prevalent methods of water harvesting, there were a few novel methods of rainwater harvesting; most prominent being the *kuin* or *par* or *teen*. It was similar to a well but the mechanism was very different.[90] It was a different kind of well in the sense that it did not depend upon underground water. In these *kuin*/wells, water would accumulate very slowly. The rainwater so retained by sand gradually percolated to the bed of the well, which was usually not more than three earthen pots in a day. In these types of wells or *kuins*, water was not always available. It took time to recharge and hence, was used for drinking purposes only. It was also perhaps known as *teen*. Nainsi informs us that the village Bhojavas of *pargana* Jodhpur had a *teen*, which could supply water twice in a day.[91] This was the technique

[90] Munhot Nainsi, *Marwar ra Pargana ri Vigat*, vol. II, ed. Narain Singh Bhati, Rajasthan Oriental Research Institute, Jodhpur, 1969, p. 96; Anupam Misra, *Rajasthan Ki Rajat Booden*, Gandhi Shanti Pratishthan, Delhi, 1995, p. 23.

[91] Munhot Nainsi, *Marwar ra Pargana ri Vigat*, vol. I, ed. Narain Singh Bhati, Rajasthan Oriental Research Institute, Jodhpur, 1968, pp. 313 and 327 and vol. II, p. 96.

which speaks of the appropriation of given environmental conditions for application of a condition-specific different method of water harvesting.

Further the chapter dwells upon the traditions of agrarian practices which have evolved in the monsoonal climate and in the semi-arid and arid landscape of the region. Methods of appropriations of possibilities and ways of mitigations have been contextualized to highlight the continuous process of adaptations. Diversity of irrigation methods and agrarian practices that evolved over a period of time to negate the uncertainties associated with the vagaries of monsoon have been enumerated. These practices bring out the strength of process of social adaptations to the climate variability.

The centrality of agrarian landscape in the history writings on medieval Indian history needs to be corrected especially when one is writing on the processes of human adaptations to the climate variability. It is pertinent on the part of any researcher examining the historical process in the monsoon ecologies of the semi-arid and arid landscape of Rajasthan. Thus, the last chapter of the book tries to capture the significance of pastoral livelihood in the sustenance of settlements in this region. Despite the prevalence of a well-established agrarian landscape, given the vagaries of climate in the region, it was difficult to sustain human settlements without making provisions for alternative modes of livelihood, that is, the pastoral component. It is argued in the chapter that human ingenuity kept its resources base wide and had avoided confinement merely to the agrarian production. Even the religious sects which flourished in the region had singularly emphasized the significance of the pastoral component.

Apparently all-pervasive, but the social manifestation/s of ecological imprint usually remains invisible. A closer examination of the social fabric of Rajasthan will enable us to decipher the ecological imprint in the different walks of life. Within the larger agro-pastoral economy of the region a closer look at the ways and means of appropriation of surplus by different sections of society will help us to decipher the diverse ecological imprints on different sections of the society. In other words, the social stratification also shapes the possibilities of resource utilization. The distribution of resources was to a great extent defined in terms of stratification. Similarly, the possibilities of production also influence the social structure. This two-way very subtle interaction has been dealt with in this volume along with illustrations of the ecological imprints in terms of social practices, rituals, etc.

MONSOON LANDSCAPE

CHAPTER 1

Intimate Visions
Environment, Territory and Society

The problem of identification of Rajasthan as a singular political, cultural or social unit continues even in the case of its geographic description. 'Geographers, for example, hold that the region "is not an object, either self-determined or nature given. It is an intellectual concept, an entity for the purpose of thought created by the selection of certain features that are relevant to an areal interest or problem".'[1] It has never been a 'natural geographic region' with clear-cut natural physical boundaries. Nevertheless, in popular memory Rajasthan has a well-defined identity in terms of cultural traits and to some extent as a broader geographic entity. Such a perception is based primarily on the literary representation of Rajasthan as a land of Rajputs who are known for their chivalry. The British colonial administrators' representation of Rajasthan as a land of 'Aryans' has played an important role in creating such an overarching identity for Rajputs and thereby, for Rajasthan as a region. Definitions of a region vary from discipline to discipline. Most definitions of a region begin with a geographical component and develop out of the relationship between the geographic features and man's adaptations to the physical environment. It is in this interaction between the landscape and man's adaptation

[1] Deryck O. Lodrick, 'Rajasthan as a Region: Myth or Reality?', in Karine Schomer, Joan L. Erdman, Deryck O. Lodrick and Lloyd I. Rudolph (eds.), *The Idea of Rajasthan*, vol. I, Manohar, Delhi, 1994, p. 2.

to the physical environment that we can trace the emergence and growth of Rajasthan, despite the presence of several well-defined sub-divisions, as a 'region' in popular perception. Nevertheless, at the outset it must be emphasized that though Rajasthan apparently represents a region, it is basically a conglomeration of numerous sub-regional identities even in terms of geographical features. Thus, it is pertinent for us to trace the development of Rajasthan as a historical entity without diluting the sub-regional peculiarities.

As we have said, it is very difficult to classify the whole of Rajasthan as a singular entity. If the Rajput states in the area shared a common history and culture, it was in the reign of the Mughals, who first drew the boundary around the region and instituted the *subah* of Ajmer, which represented the first formal delineation of Rajasthan as a politico-administrative space.

> Its length from the village of Pokhar (*Bhakar*-Pushkar) and dependencies of *Amber* to Bikaner and *Jaisalmir* is 168 *Kos*. Its breadth from the extreme limits of the *Sarkar* of *Ajmer* to *Banswarah* is 150 *Kos*. To the east lies *Agra* to the north the dependencies of *Delhi* to the south *Gujarat* to the west *Dipalpur* and *Multan*.[2]

CLIMATE

The climate of Rajasthan is characterized by an extremely high range of temperatures and aridity though it shares the characteristic variations of the monsoons through the year. It is the hottest region of India. Heat is at its peak in the western and north-western parts of Rajasthan where hot winds and dust storms are common during the day.[3] The sand dunes of the desert are the result of a dry climate with very limited or no rainfall. As far as general weather is concerned, heat dominates the scene in the whole of Rajasthan barring some hilly regions. On his way into the subcontinent Babur

[2] Abul Fazl, *Ain-i-Akbari*, vol. I, tr. H. Blochmann, Oriental Books Reprint Corporation, Delhi, 3rd edn., 1977, p. 273.

[3] Badauni Abdul Qadir, *Muntakhab-ut-Twarikh*, vol. II, tr. W.H. Lowe, Academica Asiatica, Patna, 1973, p. 239.

had realized that north-east Rajasthan is very hot.[4] Abul Fazl[5] and Jahangir[6] have placed it in the second of the seven categories in which the whole world was traditionally divided. The second category implied equitive heat and winter.

The traditional categorization of the world on the basis of climatic factors has been described by Abul Fazl:

> The learned have divided the *oikoy* into seven parts, to each of which they have given the name of *Klima*. Some reckon from the equator as Ptolemy shows in his [27] *Almagest.* Another school omitting 12° 45′ north of equator, divide the remainder and terminate as is known at 50° 31′ parallel of latitude. In the former case, therefore, the parallels from the equator will be seven circles and in the latter eight. The Seven belts which these lines form are called climates. A climate therefore is a belt on the surface of the earth between two semi-circles parallel with each other and with the equator. A climate increases in length as it approaches the equator; moreover it's first parallel will be larger than it's second.
>
> . . . There are two reasons for selection of seven as this number. The first is that ancient sages have verified by experience that each tract of superficial area was specially connected with one of the planets, as for instance, the first climate with Saturn. For this reason the inhabitants of that zone generally are dark skinned, curly haired, long-lived and indolent in action.
>
> . . . The word climate may be taken in two senses, viz., the ordinary sense in which men commonly speak of a tract of country as a climate, such as Rome, Turan, Iran and Hindustan; and the true significance already explained. In a latter meaning India is an aggregate of the first, second, third and fourth climate.[7]

The above classification very clearly highlights the traditional perception, prevalent at least during medieval times, where movements of planets were related with the changes in climate. It was common belief that planets influenced human life to a

[4] Zahirud Din Muhammad Babur, *Babur Nama*, tr. A.S. Beveridge, Oriental Books Reprint Corporation, Delhi, 1970, p. 577.

[5] Abul Fazl, *Ain-i-Akbari*, vol. II, tr. Col. H.S. Jarrett, Oriental Books Reprint Corporation, Delhi, 1978, p. 273.

[6] Jahangir, *Tuzuke-i-Jahangiri*, ed. H. Beveridge, Munshiram Manoharlal, Delhi, 1968, pp. 340–1.

[7] Abul Fazl, *Ain-i-Akbari*, vol. III, pp. 50–3.

great extent, so much so that even the colour of human skin was identified with the colour of planets. Similarly, we can infer from the above description that climates were divided along the lines parallel to the equator, without any consideration of the various geographical features, that is, rains, temperature, etc.

Francisco Pelsaert has described the region in the following words:

> The year is here divided into three seasons. In April, May, and June the heat is intolerable, and men can scarcely breathe. More than that, hot winds blow continuously, as shifting as if they came straight from the furnace of hell. The air is filled with the dust raised by violent whirlwinds from the sandy soil, making day like the darkest night that human eyes have seen or that can be grasped by the imagination. . . . The months of June, July, August, September, and October are reckoned as the rainy season, during which it sometimes rains steadily. The days are still very hot, but rain brings a pleasant and refreshing coolness. In November, December, January, February and March, it is tolerably cool, and the climate is pleasant.[8]

In the summers, the temperature rises above 40 °Celsius—the maximum occasionally reaching 50 °Celsius at places in the desert area. As Rajasthan varies a lot in physical appearance, the temperature also differs. Moreover, it is a popular belief in Rajasthan that if in the month of Ashadh, *nau-tapa* has not taken place, rains will not be good, that is, if nine days of the summer month of Ashadh (May/June) are not very hot, the cycle of the monsoon in their region shall be disturbed.[9] Rainfall also varies a lot in Rajasthan depending upon the source, route, barriers, etc. These factors combine to influence the pattern of rainfall. It varies substantially from year to year and within the same year.[10] The unequal distribution of rainfall throughout Rajasthan has been elaborated later in the chapter.

[8] Francisco Pelsaert, *Remonstrantie*, tr. W.H. Moreland and Pieter Geyl, *Jahangir's India*, W. Heffer and Sons, Cambridge, 1925, p. 47.

[9] Mishra, *Rajasthan Ki Rajat Booden*, p. 18.

[10] Concerns for variation in rainfall and its implications have been highlighted in *arzdashts* and these have been elaborated in a later section of this chapter.

VEGETATION

The impact of these varied physical environmental features of Rajasthan is most clearly visible in the nature of the vegetation. There are no large timber forests in Rajasthan, but the woodlands are extensive upon the south-western Aravallis and throughout the hilly tracts adjoining, where the rainfall is good. Mount Abu is well wooded and from Abu north-eastward the western slopes of the range are still well clothed with trees and bushes up to the neighbourhood of Merwara. The hills and ravines of Sirohi are generally wooded and some of them in the neighbourhood of Abu and Neemuch produce fine timber; while in the north and east of the state there is mostly undergrowth. All vegetation, however, rapidly decreases in the direction of Looni. In the west and south of Mewar, the forests stretch for miles.[11]

The eastern region of the state—Amber, Alwar, Mewat—had a substantial forest cover. The descriptions of the plundering acts of the Jats of the region for the latter half of Aurangzeb's reign provide glimpses of the terrain. Vakil reports have numerous references where it was stressed that due to the proximity to the forests and hills it is difficult to subjugate the Jats.[12]

The vegetation in western Rajasthan has been described in the following words:

> The Eastern and some of the southern Parganas of Marwar, in the direction of the Aravalli Range, Jaswant pura, and Jalore, are well-wooded with natural forest, and many of the ravines leading down into the plains from hills produce timber. Jaisalmer and the north and north-west of Marwar, as well as Mallani and a large sandy tract of south-west of the country towards the Runn of Kutch, produce nothing worthy of the namc of forest, although belts of Khcjra are common throughout the sandy plains, and some of the more fertile valleys in the hard desert are well stocked with these trees.[13]

[11] *The Imperial Gazetteer of India*, vol. XXI, Clarendon Press, London, 1908, p. 127.

[12] *A Descriptive List of the Vakil Reports Addressed to the Rulers of Jaipur*, vol. II (Persian), Rajasthan State Archives, Bikaner, 1972.

[13] Archibald Adams, *The Western Rajputana States: A Medico-Topographical*

The relationship between environmental features, that is, vegetation and human perception is very aptly visible in the following correlation:

> The people of Rajputana sometimes divide the country into botanical zones, the limits of which are sufficiently defined to form rough political boundaries: thus the "lai" and the "sajji' mark the desert, the "bharut" grass the country south, the "karr" grass still further south, and the "aonla" bush region marks the fertile regions of Mewar south of that.[14]

Nainsi informs us that once it was suggested that land/territory can be divided between Hardas and Sekhas on the basis of the grass it supported. Sekhas would take control of land which produced *karar* (a type of grass), while the Hardas would get the *bhurat* (a type of grass) land.[15] The above division very clearly points out that the region was identified by the kind of vegetation it supported. Even the kind of grass grown in any region became a basis of identification of land to be divided for political purposes.

Similarly, when Rao Jodha (ruler of Marwar) defeated Rana Kumbha (ruler of Mewar), it was decided that the land which sustained trees of *babool* would be bestowed to Marwar, whereas the land which sustained trees of *aanwla* would go to Mewar.[16] Once again the division as pointed out earlier very clearly highlights that the region was identified by the kind of vegetation it supported. The division of territory between states on the basis of natural environmental features once again signifies the close interaction of society with environment. The acceptance of land sustaining *babool*—signifying low productivity and arid conditions—by Rao Jodha as his territory, and territory sustaining *aanwla*—signifying a more fertile region—by Rana Kumbha substantiates the fact that

---

& General Account of Marwar, Sirohi & Jaisalmer, Vintage Books, London, 1990 (1st pub. 1899), p. 397.

[14] P.W. Powlett, *Gazetteer of the Bikaner State*, Office of the Superintendent of Government Printing, Calcutta, 1874, p. 85.

[15] Munhot Nainsi, *Munhta Nainsi ri Khyat*, vol. III, ed. Badri Prasad Sakaria, Rajasthan Oriental Research Institute, Jodhpur, 1964, p. 89.

[16] Nainsi, *Marwar-ra-Pargana-ri-Vigat*, vol. I, p. 36.

Rao Jodha controlled the arid region of Marwar and Rana Kumbha controlled the more fertile area around the Aravallis.

The above description of environmental features and their perceptions very clearly point out that Rajasthan is not a homogeneous entity. The environmental variations of the region are evident in almost every aspect of its manifestation, that is, physiographic features, climate, rains, nature of soils, river systems, etc. As a result of such a variety, the dependent flora is also varied. We have also tried to comprehend the natural vegetative growth. However, any attempt to relate these features with the human settlement, its economy, social structure, cultural practices, polity, etc., needs further delineation. The basic stress has been to highlight the possible, yet latent correlation between these features and environmental variations of the region.

LANDSCAPE

Geographically, Rajasthan presents great variability. The description of the physical features of pre-colonial Rajasthan based on contemporary sources is a difficult proposition, as we do not have any contemporary treatise which comprehensively deals with the geography of the entire region. However, since geographical features do not change overnight, we can supplement our contemporary evidences, both archival and non-archival, by making use of even nineteenth-century records, such as the gazetteers, travelogues, and other non-archival literature, etc.

Similarly, it is necessary to point out that the portrayal of even the environmental features in various sources changes along with the respective perceptions of social needs, political concerns, cultural traits, etc. Ann Grodzins Gold has aptly argued: 'One important observation to emerge is that the ways that elements of nature have been viewed and treated in different eras and regions are highly variable according to local political and ecological specificities.' Further, she stresses:

> Elements of the environment become emblems of satisfaction and deprivation, submission and confrontation. One person recalls blistering his feet in the dry riverbed on a frivolous errand for the king; another

remembers the exquisite thrill and dire risk of poaching and consuming savory wild boar.[17]

Thus, the representations of environmental features in contemporary and near contemporary sources also differ. The locus standi of various sources influenced the portrayal of environmental features. Along with epigraphic sources, literary works like Munhot Nainsi's account, *arhsattas*, *Jaisalmer ri Khyat*, *Vir-Vinod*, *Twarikh-i-Jaisalmer*, etc., basically cater to politico-administrative needs. Hence, these records provide a detailed description of physical features important primarily for production purposes, especially agricultural production. There are another set of accounts, like that of Tod's travelogue, where orientalism was romanticized. Tod's attitude to Rajput society laid emphasis on two main themes—heroism and love. His Rajput heroes are honourable in love and war while his heroines are chaste.[18] Environmental features were described to highlight the adverse conditions which were faced by these heroes. Sources thus paint different images for the same region.

In popular perceptions today, Rajasthan attains the identity of a desert landscape, however wrong it may be. This image of Rajasthan also has a corollary in the image of the 'Rajput'. The 'Rajputs' as warriors could fight against the Muslim state of the Mughals because they have been imagined as hardy and those who could survive the harshness of desert conditions. The construction of such an image appears to be a very recent phenomenon[19] since sources from the past offer a very different image of human–nature interaction. Nevertheless, as pointed out earlier, the emergence of Rajasthan as a 'region' can be traced to the establishment of Mughal rule in north India. Its subsequent organization as a politico-administrative region under the Britishers led to further

[17] Gold and Gujar, *In The Times of Trees*, pp. 12–14.

[18] Rajat Kant Ray, 'Colonel James Tod, Muhanota Nainasi and the Rajputs', *Indian Historical Review*, vol. XXV, no. 2, June 1999, p. 101.

[19] Dirk H.A. Kolff, *Naukar, Rajputs and Sepoy: The Ethnohistory of the Military Labour Market in Hindustan, 1450-1850*, Cambridge University Press, Cambridge, 1990.

consolidation of this identity, which is still in vogue. In the British period 'Rajputana' was an administrative nomenclature for a great historical circle, which included eighteen native states, two chieftaincies and the British district of Ajmer–Merwara. However, it was James Tod who used the expression in a larger and wider context, defining Rajputana as 'that part of the country where Rajput rulers reside'.[20]

Unlike the unifying factor proposed by Tod, Rajasthan can be geographically and climatically divided into several diverse niches. The earliest geographic reference to Rajasthan is in connection with its aridity. The word *maru* is found in the *Rgveda* (1.35.6).[21] However, Rajasthan cannot be summarized by this term as *Maru Desha* stood specifically only for the desert area of Rajasthan. In this context, it is important to keep in mind the fact that as far as the Rajasthan desert is concerned, 'scientists conclude that the northern boundary of desert never remained in fixate'.[22] Similarly, the term *jangal*[23] has also been used, but to describe specifically the region of Bikaner. 'From the view point of vegetation, the word *Jangal* denotes a region where sky is generally clear and water are scarce, and land abounds in *Sami* (Prosois Spicigera), *Karira* (Capprisaphyylla), *Pilu* (Carya Rborea) and *Kankandhu* (Ciziphus Jujuba) trees.'[24]

The Aravalli range of hills running diagonally from south-west to north-east divides Rajasthan broadly into two natural divisions, namely, the north-western, which is arid, and the south-eastern, being the semi-arid. About three-fifths of Rajputana lies north-west of this line leaving two-fifths on the south-east. The south-western end of this range is known as Mount Abu, whereas it subsides near Shekhawati in the north-east, forming a great natural rampart

[20] Tod, *Anals and Antiquities*, vol. I, p. 1.

[21] D. Sharma, *Rajasthan Through the Ages*, Rajasthan State Archives, Bikaner, 1966, p. 11.

[22] G.S.L. Devra, 'Desertification and Problem of Delimitation of Rajputana Desert During the Medieval Period', *Human Ecology*, Special Issue, no. 7, 1999, p. 97.

[23] Zimmerman, *Jungle and the Aroma of Meats.*

[24] Sharma, *Rajasthan Through the Ages*, p. 12.

across the entire length. Its continuation can be traced into the low ridges of Satalmer and Jaisalmer. Even these two regions are not homogeneous entities. The variations are visible at several levels, that is, physical forms, geological features, climate, river system, nature of vegetation, etc.

Among the various categories, the first two, that is, the physical forms and geological features are more or less static whereas the remaining are more volatile, prone to external disturbances/ influences. Even in their volatile temperament these generally represent a periodicity. These are manifested in the form of seasons. The most prominent of these has been the monsoon, which originates in the distant seas. These monsoon winds have to travel a long distance before reaching Rajasthan. In the process, the physiographic features of regions thus traversed become significant. Therefore, before venturing further into Rajasthan it is pertinent for us to situate Rajasthan in the wider geographical context, that is, the pre-partition Indian subcontinent.

Rajasthan is bound on the west and the north-west by the Thar Desert, which is known as the Great Indian Desert and politically comprises the province of Sindh and Bahawalpur. On the north and the north-east, we have the very fertile region of Punjab irrigated by numerous tributaries of the Indus—Sutlej, Ravi, Beas, Jhelum, etc.—and beyond that stretches the Great Himalayan mountain range. The eastern frontier of Rajasthan is part of the Gangetic Doab of Agra and Oudh, followed by the hilly terrain of Gwalior sustaining various tributaries of the Jamuna, that is, Chambal, Betwa, Banas, etc. The southern boundary runs across the central region of India, which is interspersed by the mountain ranges of the Vindhyas and Satpura. The south-western region shares boundaries with the Rann of Kutch.

Similarly, the vast arid extent of western Rajasthan had also functioned as a formidable barrier to any attempt at territorial expansion made by the tribes located in the middle-east, that is, Afghanistan and beyond. Even the Britishers were aware of this geographical barrier. 'On the north-west lies the Thar, the great Desert of India, which from time immemorial has proved to be a more effective barrier to the advance of armies than the Indus itself.'[25]

Rajasthan may be divided into three geological regions: The central and the largest region comprises the whole width of the Aravalli system, formed of very old sub-metamorphic and gneiss rocks. The second is the eastern region with a sharply defined boundary, along which the most ancient formations are abruptly replaced by the great basin of the Vindhyan strata, or are overlaid by the still more extensive spread of the Deccan trap, forming the plateau of Malwa. The third is the western region, of very ill-defined margin, in which, besides some rocks of indeterminate age, it is known or suspected that tertiary and secondary strata stretch across from Sindh, beneath the sands of the desert, towards the flanks of the Aravallis. Rajasthan may be considered rich at least in the variety of metals it produces. Zinc[26] is extracted from this region. Copper and lead exist in several parts of the Aravalli range and in the minor ridges of Alwar and Shekhawati. Iron ore abounds in Alwar, Mewar and Jhalawar.

The river systems of Rajasthan also vary a lot like its physical features. It is the physical features—contours, climate, etc., that usually determine the river system of any region. There are two distinct slopes, one towards the south-east and the other towards the north-east, with the Aravalli mountain ranges forming the watershed in-between these two slopes. These hills of Rajasthan[27] are the source of most of the water courses. All this water is drained towards the south-east into the Banas. The north-eastern slope has a river system of importance supported by river Chambal.

The Chambal,[28] being the longest river, flows through the territory for about one-third of its course, and forms its boundary for another third. The Banas, which is next in importance to the

[25] *Imperial Gazetteer of India*, vol. I, Clarendon Press, Oxford, 1907, p. 33.

[26] Shayamal Das, *Vir Vinod*, vol. I, Motilal Banarsidass, Delhi, 1986, pp. 108–9; Tod, *Annals and Antiquities*, vol. II, p. 505.

[27] Babur, *Babur Nama*, p. 486.

[28] Ibid., p. 485.

Chambal, rises in the south-west near Kankraoli in Mewar. It collects nearly all the drainage of the Mewar Plateau, with that of the south-eastern slopes and hill tracts of the Aravallis.[29] It joins the Chambal a little beyond the north-eastern extremity of the state of Bundi, after a course of about 300 miles. Among the south-western hills of Mewar, the western Banas[30] and the Sabarmati take their rise, but attain no size or importance until after passing the Rajputana frontier towards the south-west.

The south-western part of Rajasthan is a desert with Looni being the only river of importance. It rises in the Pushkar Valley close to Ajmer, and runs south-west for about 200 miles into the Rann of Kutch.[31] It receives and cuts off from the western plains all the drainage brought by the mountain torrents down the western slopes of the Aravallis between Ajmer and Abu. Beyond Looni, to its north-west and north-east, there are no perennial streams.

The Mahi,[32] a considerable river in Gujarat, runs for some distance through the territories of Pratapgarh and Banswara, in the southern part of the region. Its chief tributary in this part is the Som,[33] which run first east and then southward through Mewar. These rivers carry off the drainage of the south-west corner of Rajasthan into the Gulf of Kutch and Cambay. Rajasthan has no natural fresh water lakes, the only considerable basin being the well-known salt lake at Sambhar. There are some artificial lakes in Mewar and numerous tanks and *talabs* spread all over the region.

RAINFALL PATTERNS

Rajasthan probably presents a greater variety of climatic conditions and factors than any area of similar size in the world. The physiographic and geographical features of Rajasthan are of great importance, in so far as they modify more or less considerably,

[29] Ibid.

[30] Das, *Vir Vinod*, p. 103.

[31] *Imperial Gazetteer of India*, vol. XXI, p. 86.

[32] Munhot Nainsi, *Munhot Nainsi ri Khyat*, vol. I, ed. Badri Prasad Sakaria, Rajasthan Oriental Research Institute, Jodhpur, 1960, p. 34.

[33] Ibid.

the lower air movement and hence, humidity and rainfall. It is thus necessary to take note of these factors while attempting any description of the environmental features of Rajasthan.

As discussed above, the western side of the region comes very near the limits of that part of Asia that belongs to the rainless district of the world; though even on this side, the south-western winds bring a little rain annually from the Indian Ocean. In this western part—Jaisalmer, Bikaner and the greater part of Marwar—the annual rainfall scarcely averages more than 5 inches. This is because the rain clouds have to pass extensive and heated tracts of sand before reaching these plains, and are emptied of much of their moisture upon the high ranges in Kathiawar and the nearer slopes of the Aravallis.[34]

In the south-west which is more directly accessed by the periodical rains and with less intermediate evaporation, the rainfall is much more copious; and at Abu it sometimes exceeds 100 inches. However, except in these south-west highlands of the Aravallis, the rain is most abundant in south-east Rajasthan. The southern states, from Banswara to Jhalawar and Kota, get not only the rains from the Indian Ocean which sweep up the valleys of the Narmada and Mahi rivers across Malwa to the countries around the Chambal. The region also receives the last of the rains which come up from the Bay of Bengal in the south-east; and this supply occasionally reaches all Mewar. In this part of the country, if the south-west rains fail early, the south-east rains usually come to the rescue later in the season; so that the country is rarely subjected to the extreme droughts of the north-western tracts.[35]

In the central part of Rajasthan around Ajmer towards Jaipur, the periodical supply of rains is very variable. If the eastern winds are strong, they bring hard rains from the Bay of Bengal; whereas if the south-west monsoon prevails the rain is comparatively late and light. Sometimes a good supply comes from both seas, and then the rainfall is more; but it is usually much less. In the far north

[34] *The Rajputana Gazetteer*, vol. I, Office of the Superintendent of Government Printing, Calcutta, 1879, p. 20.

[35] *The Imperial Gazetteer of India*, vol. XXI, p. 92.

of Rajasthan, the winds must be very strong and the clouds very full to bring any appreciable supply from either direction.[36] The northern part of Rajasthan gets a scanty share of the winter rains of north India, while the southern parts usually gets none at all, beyond a few soft showers around Christmas. It may be said that from Bikaner and Jaisalmer in the north-west to Pratapgarh and Kota in the south-east, the rainfall very gradually increases from 5 to about 45 inches, the quantity increasing very rapidly after the Aravallis have been crossed.[37]

CULTURAL CONSTRUCTS OF REGION AND IDENTITIES

As pointed out earlier, Rajasthan has massive diversity in terms of environmental features. Conventional descriptions of Rajasthan are based on two broad divisions. The first comprises the west and north-western region beyond Aravallis, the second covers the south-eastern region. The *dhundhar* (Amber–Shekhawati) region has overlapping features of both the regions. However, it is to be pointed out that in terms of environmental features, even these two broad divisions are not homogeneous entities. Therefore, we propose to base our description on the traditional political divisions of Rajasthan. These in turn would help us explore the possible co-relation between physical geography and the resultant social formations.

Looking first at the arid division of Rajasthan that lies north-west of the Aravallis, we find a vast tract stretching from Sindh on the west along the southern Punjab frontier to the outskirts of Delhi on the north-east. On the whole this segment is sandy,[38] ill-watered and unproductive. The situation improves as one moves towards the north and east. Sindh lies on the west of Rajasthan in the Great Desert. The low ridges of limestone, which lie east of the Great Desert, protect the area against desert winds. Further

[36] *The Rajputana Gazetteer*, vol. I, p. 20.

[37] Ibid., pp. 92–3.

[38] Ibid., pp. 182–219; Jean Baptiste Tavernier, *Travels in India*, tr. V. Ball, Oriental Books Reprint Corporation, New Delhi, 1977, p. 63.

eastward is the 'Little Desert', which runs up from the river Looni between Jaisalmer and Jodhpur, extending towards the north-east up to Bikaner.

The character of the desert region is almost the same everywhere. It is also known as Thar.[39] It is covered with sand dunes, shaped generally in long, straight ridges, which seldom meet, but run in parallel lines, separated at short and fairly regular intervals, resembling something like the ripple marks on the seashore upon a very magnified scale. Some of these ridges may be 2 miles long, varying from 50 to 100 ft in height. Their sides are well marked and from a distance they look like low hills. Their summits are blown up and carved like waves due to the action of the periodical westerly winds.[40] They are sparsely covered with stunted shrubs and tufts of coarse grass in the dry season, and the light rains cover them with vegetation.

Yuan Chwang, passing through the desert region in AD 643 describes the region around Bhinmal (Pi-Lo-mo-lo) as a place where the climate was hot, and there was much wind and dust, and the soil was sandy and fruits and flowers were rare.[41] Somewhat similar features are described in the literary sources. Contemporary literary representations of the region include the following:

थली वर्णन

थली सिरदार वर्णन

दसे दिस चावउ भोलो देस । उंफडा जल पीवइ बुरि असेस ।।
कहइ जे बोल मिलई सुखकार । सहु सिरि देस थली सिरदार ।।
जिहां नही कूड़ कलेस बजार । गिणही नहिं किणसूँ द्रोह लिगार ।।
धन थोड़े धर करइ साधर । सहु सिरि देस थली सिरदार ।।
धरा जिण बालक धर्म सु धीर । आया जिण काज अखंड अभीर ।।
नियाहण टेक जिहां नर नार । सहु सिरि देस थली सिरदार ।।

[39] Perhaps the term 'Thar' is derived from *thal,* which means desert. Munhot Nainsi, *Marwar-ra-Pargana-ri-Vigat*, vol. III, Narain Singh Bhati (ed.), Rajasthan Oriental Research Institute, Jodhpur, 1974, p. 132; Sita Ram Lalas, *Rajasthani Hindi Sabad Kosh,* vol. I, Rajasthani Shodh Sansthan, Jodhpur, 1986, p. 617.

[40] Sharma, *Rajasthan Through the Ages*, p. 45.

[41] T. Watters, *On Yuan Chwang Travels in India*, vol. II, Munshiram Manoharlal, New Delhi, 1973, pp. 243–50.

करइ चउकस जे होवै काम । सदा परदेस बीह जू ठाम ।।
करइ जिहाँ लोग सदा उपगार। सहु सिरि देस थली सिरदार ।।
जिहां नहीं रोग संभावइ जोर। भुजगल महा बलवंत अभोर ।।
डोलाला डाकी नइ डोकार। सहूसिरि देस थली सिरदार ।।
जिसा थल जोर अजंग अजेय । भला जिहां लोक गंभीर अभेय ।।
नर नारी होइ रतन्न न बार । सहु सिरि देस थली सिरदार ।।
जिहां भय होइ न दंड न जोर । चिहुं दिसि चावउ कोइ न चोर।
गटका जिहां दूध दही उद्‌गार। सहु सिरि देस थली सिरदार ।[42]

It is the best country in the world. Water is available at great depth, but is very good for the mind. It is very famous, and everybody knows, that *Thali is the king* of the land. There is no trouble on this land. Nobody fights. Prosperity is limited, however, people are religious. *Thali* is the king of the land. Nobody talks ill of anybody, people are soft spoken and they believe in truth. *Thali* is the king of the land. Even the children believe in *Dharma*—righteousness. People will help you in the path of religion. People are helpful. People will do work with perfection. People from other regions come and settle here. People help each other. *Thali is the king* of the land. Here epidemics will not take place. Young men play with swords and even the old ones are very agile. *Thali* is the king of the land. The *thal* has been unconquered because the public is brave. People are like gems. *Thali is the king* of the land. People believe in charity. Even a little donation is enough to make one famous. People worship their teachers. *Thali is the king* of the land. There is nothing to fear or any need to use force. Honesty is everywhere; there is no theft. The land is full of milk and curd. *Thali is the king* of the land.

थली दोस वर्णन

उड़ई जिहां खेह न थंभि रहइ । वज्जई जिहां पवन न किउही सहइ ।।
जल खारउ सोइ पावेइ वली । पिफ्ट देस कुदेस कुखंड थली ।।
पट मासे नीर निवाण लहइं । जिहां चउपद जीवति स्याइ रहइ ।।
जिहां त्रास जलइ नर आस पफली । पिफ्ट देस कुदेस कुखंड थली ।।
जिहां सूरख लोग पिसाच जिसा । काला अति भूछ कि भूत जिसा ।।
भरी रीवड छाछि पिवंति रली । पिफ्ट देस कुदेस कुखंड थली ।।
जिहां डाढी मूछ न सूढ किसी । मिली भूहन दीसइ मुख दिसी ।।

42 Agarchand Nahata (ed.), 'Thali Varnan', *Maru Bharti*, Year II, no. 1, 1954, pp. 81–3.

जम्मारउ जाबइ साथि हली । पिफ्ट देस कुदेस कुखंड थली ।।
जिहां तरबर सांगर बोर तणा । तिणि पाखलि दीसई साप घणा ।।
मिलि तोड़इ क्राढ़ै लोक लली । पिफ्ट देस कुदेस कुखंड थली ।।
उफघाड़इ माथई कूक करइ । चड़ि खेजड़ि सांगर भार भरइ ।।
दीसंता भूत कि दैत दली । पिफ्ट देस कुदेस कुखंड थली ।।
नित नीर न लागइ नीर जिया । भलउ भख्य न देखइ जेथ तिया ।।
जिहां सुख न कोई भेद मिली । पिफ्ट देस कुदेस कुखंड थली ।।
जिहां कांटा क्रँपड़ वाडि करइं । गाडां धरि हांडल तेथ भरइ ।।
पसु जीवित जीवई कवण कली । पिफ्ट देस कुदेस कुखंड थली ।।
जल जेध न पावई पक्षी जिहां । तरू छेह न दीसइ पंथि तिहां ।।
चढ़ि खेजड़ क्रपड़ि सार चली । पिफ्ट देस कुदेस कुखंड थली ।।
धरा थल पंथी दुक्ख धरइ । भरि पाए लागि भुरट्ट भरइ ।।
जिहां बोलै बसती अेह जली । पिफ्ट देस कुदेस कुखंड थली ।।
जिहां पीवइ सांढां दूध सकोइ । मरि मोगा ठामो ठामइ हाइ ।।
लागइ उठि पंथी लोक बली । पिफ्ट देस कुदेस कुखंड थली ।।
जिहां पीवण सांप निसा विचरइ । मणिधर पन्नग पफोंकार करइं ।।
धरि धवल देह आनंद रली । पिफ्ट देस कुदेस कुखंड थली ।।
जिहां छक्कड़ दुक्कड़ द्रव्य घणाँ । जिहां जउवाकेरी कांई नमणा ।।
जिहां पूफल न पान न तेल पफली। पिफ्ट देस कुदेस कुखंड थली ।।
जिहां कांटा आगलि पंथ नहीं । कोइ मानव पंथि न दीसई जहीं ।।
दादउ तिहां पावइ नीर मिली । पिफ्ट देस कुदेस कुखंड थली ।।
जहां ढोल दमक्कइं नीर भरइ । साठीका कोहर कूक करइं ।।
केई नारी उफपरी नारि ढलो । पिफ्ट देस कुदेस कुखंड थली ।।
जिहां उफन उखेली वस्त्रा करइं । धण माडर छाली दूध धरइं ।।
उफधड़ै गाडै वास गली । पिफ्ट देस कुदेस कुखंड थली ।।
जिहां काचर बोर घमंड करइ । बलि रांध मतीरा पेट भरइ ।।
मन मांहे आयो आप रली । पिफ्ट देस कुदेस कुखंड थली ।।
जिहां उफति चढी दिन राति बहइ । दस कोसां अेको कोस कहइ ।।
जिहां आकां कयरा छांह छली । पिफ्ट देस कुदेस कुखंड थली ।।
कहुं केता थली तणां बाखांण । अे माणस नांही छइ पाखांण ।।
कह्या गुण देखी किउं संभली । पिफ्ट देस कुदेस कुखंड थली ।।[43]

[43] Ibid.

It is a place where sand blows without any restriction. Here wind blows at great speed. The water is brackish, so it is not potable hence *thali* is such a bad land to live in. Ponds retain water only for six months, and livestock suffers due to limited availability of water. *Thali* is such a bad land to live in. Here people are like *Pishach*—ghosts and as black as ghosts. People drink a lot of *chhachh. Thali* is such a bad land to live in. People keep beard and moustache and their face is full of dust, which blows here continuously. It is as if *yamraj*, the god of death, accompanies the peasants who work in these conditions. *Thali* is such a bad land to live in. Here *Sangri*—dominates the vegetation, which is an abode of snakes. . . . *Thali* is such a bad land to live in. People speak very loudly. They fetch *Sangri* from the trees of *Khejri.* People look like ghosts or demon. *Thali* is such a bad land to live in. People cannot take bath daily, not much is available as food and there is no pleasure. *Thali* is such a bad land to live in. Here vegetation is full of thorns. Here water is not available for birds, trees do not provide shade and *khejar* dominates the land. *Thali* is such a bad land to live in. It is very difficult for someone to travel in these conditions as *buhrat* grass pierces the feet. Heat makes it difficult to live even in the villages. *Thali* is such a bad land to live in. One gets milk of camel to drink. People migrate to other places. *Thali* is such a bad land to live in. Here yellow snakes roam around in the night. *Manidhari*—snakes with gems on their forehead play in the night. There are plenty of cattle and carts, which help in trade. There are no flowers or greenery. *Thali* is such a bad land to live in. There is no road which is not full of thorns. Even on long stretches of road there is no traveller. *Thali* is such a bad land to live in. The wells are so deep that water can be fetched only with the help of camels/bullocks. Water being so deep the fetching rope has to be very long. A camel is used to draw the rope up. In the process the camel goes so far that when the bucket comes up, it is informed by the beating of a drum. Such wells are known as *saathi ka kuwa. Thali* is such a bad land to live in. Here cloths are woven from the wool, although milk is available in plenty, it is mainly camel and goat milk, grass is produced in abundance. *Thali* is such a bad land to live in. We have only *kachar, Ber*, they make vegetable of *matira.* These are the staple diet of *thali. Thali* is such a bad land to live in. This is the place where day reigns supreme and nights are short. Land is so destitute that people call 10 *kos* of distance as only one. During day time whatever shade is available is from the bushes of *aak. Thali* is such a bad land to live in. This is how *thali* is and it is not an exaggeration. These are its qualities also. *Thali* is such a bad land to live in.

An interesting feature of contemporary literature is the almost simultaneous mention of both the virtues and vices of the region. There are various reasons to settle in the region described as the best country in the world. The water is available at great depths, but is very good for the mind. Prosperity is limited but people are religious. People are helpful and hard working. People from other regions come and settle in this epidemic-free region. Young men are playful and even the old ones are very agile. Similarly, the vices have also been described at great length. Sand blows unhindered in this region. Wind blows here at a great speed. The water is brackish, so it is not potable. Ponds retain water only for six months, and livestock suffers due to limited availability of water.

Boileau also shared a similar experience. He describes it as:

our journey from the Shekhawat frontier to Poogul, a distance of 180 miles, was over hills and valley of loose and heavy sand. The hills were exactly like those which are sometimes formed by the wind on the seashore, but far exceeding them in height, which was from 20 to 100 feet. They are said to shift their position and to alter their shapes, depending on the effect of wind; and in summer, the passage of many parts of the desert are said to be rendered dangerous by the clouds of moving sand; but when I saw the hills (in winter) they seemed to have a great degree of permanence, for they bore a kind of grass, besides *phok* (*phog*), and the thorny branches of the *babool* and the *bair* or *jujuba*, which altogether gave them an appearance that sometimes amounted to verdure. Among the most dismal hills of sand, one occasionally meets with a village, if such a name can be given to a few round huts of straw with long walls and conical roofs like little stacks of corn. . . . These miserable abodes are surrounded by hedges of thorny branches stuck in the sand. . . .'[44]

Further, he recounts that:

If I could present to my reader the foreground of high sand hills, the village of straw huts, the clay walls of the little fort going to ruins as the ground which supported them was blown away by the wind, and sea of sand without a sign of vegetation which formed the rest of the prospect, he probably would feel, as I did, a sort of wonder at the people who could

[44] A.H.E. Boileau, *Personal Narrative of Tour through the Western States of Rajwara in 1835 Comprising Beekaner, Jesulmer & Jodhpur*, Baptist Mission Press, Calcutta, 1837, pp. 43–7.

reside in so dismal a wilderness, and horror of the life to which they seem to be condemned.[45]

However, the situation was not as dismal as painted by Boileau. Perhaps he has tried to compare the greenery of his homeland with the aridity of Rajasthan. The people living in the region had adapted themselves to the climate. This is reflected in the conversation between peoples living in different regions, wherein every speaker highlights the beauty of one's own region while criticising other regions' problems. It is very much evident from the description in the folktale of *Dhola Maru Ra Duha* where Malwani, being a resident of Malwa, points out the problems with the environment of Marwar, whereas Maruni, being a resident of Marwar, defends Marwar:

ततखण माळवाणी कहइ, साँभळि कंत सुरंग । सगळा देस सुहाँमणा, मारू-देस विरंग ।।
बाळउँ, बाबा, देसड़उ, पाँणी जिहाँ कुवाँह । आधीरात कुहक्कड़ा, ज्यउँ माणसाँ मुवाँव ।।
बाळउँ, बाबा, देसड़उ, पाँणी-संदी ताति । पाणी-केरइ कारणइ प्री छंडइ अधराति ।।
बाळँ, ढोला, देसड़उ जइँ पाँणी कूँवेण। कूँ-कूँ-वरणा हथ्थड़ा नही सुँ घाढा जेंण ।।
बाबा, म देइस मारूवाँ, सूध एवाळाँह । कंधि कुहाड़उ, सिरि घड़उ, वासउ मंझि थळाँह ।।
बाबा, म देइस मारूवाँ, वर कूँआरि रहेसि । हाथि कचोळउ, सिरि घड़उ सीचंती य मरेसि ।।
मारू, थाँकइ देसड़इ एक न भाजइ रिडड । ऊचाळउ क असरसणउ, कई फाकउ, कइ तिडड ।।
जिण भुइ पन्नग पीयण, कयर-कँटाळा रूँख । आके-फोगे छाँहड़ी, हूँछाँ भाँजइ भूख ।।
पहिरण-ओढण कंबळा, साठे पुरिसे नीर । आपण लोक उभाँखरा, गाडर-छाळी खीर ।।
मारू-देस उपन्निया तिहाँका दंत सुसेत । कूझ-बची-गोरंगियाँ, खंजर जेहा नेत ।।

[45] Ibid.

मारू-देस उपन्निया, सर ज्यउँ पध्धरियाह । कड़वा कदे न बोलही, मीठा बोलणियाह ।।
देस निवाणूँ, सजळ जळ, मीठा-बोला लोइ । मारू कॉमिणि दिखणि धर हरि दीयइ तउ होइ ।।
देस सुरंगउ, भुइँ निजळ, न दियाँ दोस थळाँह । घरि-घरि चंद-वदन्नियाँ, नीर चढइ कमळाँह ।।[46]

Malwani says that except Marwar, all other regions are good. She complains that in Marwar it is so tedious to fetch water from deep wells that people begin this chore from mid-night itself. She says I would not live in such a region where husbands leave home at midnight to fetch water while the wives cannot do so as the excessive depth of water in wells makes it a laborious process. I wish that my father would not marry me off in Marwar because it is a place of pastoralists; one has to carry a vessel to fetch water and an axe. They live in the desert. I would prefer to stay unmarried than to go to Marwar where hands are bruised trying to fetch water from the deep wells. She tells Marwani that Marwar is full of problems. There is either drought or rains are delayed or one has to migrate or a locust attack. In Marwar, snakes abound and in the name of vegetation one gets only grass of Kair and Katra are considered as tree and bushes of Fog and Akra is the source of shelter from the sun. It is a place where water is available in wells only at a depth of sixty arms' lengths. Due to limited agricultural production, people are nomadic and get milk of sheep and goat only.

To these criticisms Maruni answers that Malwa is not a good place to live and defends Marwar. She says that people of Marwar are fair coloured and their teeth are bright. People of Marwar are soft spoken and straightforward. Marwar is full of tanks and the water of the region is good for health. She says that a girl from Marwar and the state of Narwar is bestowed only on lucky persons.

The previous section points out that the common man identifies with the environmental features of the place of his residence. The environmental features are put to better use to earn a living and to sustain the family. The hardships of the region are perceived as an asset and they do not want to migrate and settle in other

[46] Mahavir Singh Gahlot (ed.), *Dhola Maru Ra Duha*, Rajasthani Granthagar, Jodhpur, 1985, pp. 176-80.

regions. Such opinions expressed in the literature capture not only the popular attachment to one's lands but also the fact that they are willing to adapt and find ways and means to deal with the hardships, instead of migrating to greener pastures and areas of greater geographical comfort.

The people were well aware of the severe nature of the climate of their region. It is very much evident from the analysis of one of the folk songs dealing with women collected by Bhoju Ram Gujar. While describing an episode of the life history of the regional *avatar* of god Vishnu, Dev Narayanji, it is said, 'When Dev Narayan took birth on Malaseri hill, that place became Mathura City. From happiness the dry tree became completely green and the peacocks' voices began to cry.'[47] The striking contrast between the desolate and arid Rajasthan on the one hand and the green surroundings of Mathura on the other is clearly brought out.

To further delineate the environmental features we shall broadly take recourse to the traditional (sixteenth-eighteenth century) political divisions of the region. The westernmost part of this arid region is known as Jaisalmer. It is almost entirely a sandy waste,[48] forming part of what is called 'the Great Indian Desert'. In the neighbourhood of Jaisalmer city the soil is very stony. One encounters sandstone rocks, flat topped and entirely destitute of vegetation cover; but with this exception the general aspect of the country is that of an interminable sea of sand hills of all shapes and sizes, some rising even to a height of 150 ft.[49] The arid nature of this region has been aptly summarized by Nainsi. Writing of the details of a place called *Mangli ka Thal* (west of Jaisalmer), he says:

तिकै जैसळमेरथा कोस २५ आथवणानूं मंगळका-थळ छै, तडै रहै छै । बा ठोड़

[47] Ann Grodzins Gold and Bhoju Ram Gujar, 'Of Gods Trees & Boundaries', in N.K. Singhi and Rajendra Joshi (eds.), *Folk, Faith & Feudalism*, Rawat Publications, New Delhi, 1995, p. 40.

[48] Thevenot says, '. . . countries of Gesselmere and Bando, are Barren . . .', Surendra Nath Sen (ed.), *Indian Travels of Thevenot and Careri*, National Archives of India, New Delhi, 1949, pp. 73–4.

[49] Shayamal Das, *Vir Vinod*, vol. II, pt. III, Motilal Banarsidass, Delhi, 1986, pp. 1751-3.

> मंगळीका-थल कहावै छै। तठै द्रम छै । सु भोमियो होय सु डांडी आवै । असैंधौ डांडो टळै सु घोड़ो असवार गरह हुजाय ।[50]

The region is full of dust storms. It is such a desolate desert that if a newcomer misses the path, he shall be lost forever and die.

The land in the western portion of the state is covered with *phog*[51] bushes (calliganum), while that in the eastern part supports tufts of long grass. The trees of *jal* (*peelu*) and *khejari*[52] are reported from this region.[53] In the east near the large villages of Noh, Bikampur and Birslipur, fields have been formed in the valleys between the sand hills, where when the season is favourable the inhabitants can grow *jawar*[54] and *bajra*. The settlements are generally situated near sources of water.[55] The Karad region of Jaisalmer has seventeen villages with the provision of water in the form of *talabs*/*talai*, wells.[56]

> कोटड़ी, छ्होटणरा भाखरांरो पांणी आवै तिणसूं भरीजै, पाखती च्यांरु तरफ भाखर छै, नै बीच उंफडाळ छै । कोस ३ बीच पांणीसूं भरीजै तद दस पनरै बांस पांणी चढै । पांणी निकलणरी ठोड़ को नहीं । सबळो भरीजैं तद हासल इजाफ हुबैय । पांणी घटै तद मांहै बेरी दोय सौ, च्यारसौ आखारी सी हुबै छै ।[57]

The settlement is enclosed by hills from all sides. There is a depression in the middle of the hills, which have enclosed it. The runoff water of the enclosing mountains is collected in the depression, as there is no outlet. When the water is full it spreads up to 3 *kos* and is 10 to 15 poles deep. When it dries up it provides potential for a good harvest in its bed and it also sustains two to four hundred wells in its bed.

[50] Nainsi, *Munhot Nainsi ri Khyat*, vol. II, p. 31.

[51] *Impereial Gazetteer of India*, vol. XIV, Clarendon Press, London, 1908, p. 1.

[52] Lakmi Chand, *Tawarikh Jaisalmer*, Rajasthani Granthagar, Jodhpur (1899), rpt. 1999, pp. 215-16.

[53] Nainsi, *Munhot Nainsi ri Khyat*, vol. III, p. 108.

[54] Ibid., vol. II, p. 5.

[55] Ibid., p. 99.

[56] Ibid., p. 12.

[57] Ibid., p. 5.

Water is scarce, and generally brackish. The average depth of wells varies between 156 to 350 ft.[58] Rainwater is used for drinking purpose as far as possible; but owing to the scanty rainfall, the supply stored up in the village dries up as summer begins, forcing the dependent population to look for another source of water. At times it leads to temporary migration.

There are no perennial streams in Jaisalmer,[59] and only one small river, called the Kakani.[60] It rises near the villages of Kotri, Gohria and Latabra and after flowing a distance of 28 miles, spreads over a large surface of flat ground, forming a depression/lake or *khabha*, near Buj.[61] There are a few other similar rivers, that is, Lathi, Mogari, Sookari, Bankia, etc., which are seasonal.[62] Except in the years 1849, then 1852, and the latest in 1938, River Lathi could seldom run its full course. The possible reason for its failure has been suggested to be the construction of several embankments and the formation of new sand dunes.[63]

On the south and east of Jaisalmer we have Marwar, another important division of the arid region. River Looni is the most marked feature in the physical aspect of Marwar. It rises in the lake of Ajmer[64] and is first known as the Sagar Mati taking the name of Looni after its junction at Govindgarh with the Sarsati. From Govindgarh the river flows in a south-westerly direction through the region and is finally lost in the marshy ground at the head of the Rann of Kutch. Numerous tributaries, chiefly from the Aravalli Hills, feed this river. The chief tributaries of the Looni are the Jagirs, the Sukni, the Reria or Pali, the Banda and the Juwai.

The river is, however, capricious and erratic. On one bank it may be a blessing, on the other a curse. The Looni attains its greatest breadth in the Sanchor and Mallani districts. During summers

[58] Adams, *The Western Rajputana States*, p. 123.

[59] Chand, *Tawarikh Jaisalmer*, p. 205.

[60] Ibid.; जैसलमेर था कोस ७ आथूणनूं काक नदीरो पांणी आवै । Nainsi, *Munhot Nainsi ri Khyat*, vol. II, p. 5.

[61] Chand, *Tawarikh Jaisalmer*, p. 205.

[62] Ibid., pp. 206–7.

[63] Ibid.

[64] *Imperial Gazetteer of India*, vol. V, Clarendon Press, London, 1908, p. 138.

Looni contains only sandy pools of water and its tributaries are dry during ten months of the year. There is a saying in Marwar, that 'half the produce of the country, so far as cereals are concerned, is dependent on the Looni'. It runs between the sand and the banks are very low, allowing overflow so that water spreads to a larger area along the banks. It spreads to almost 5 miles in breadth, leaving as it recedes, a rich alluvial soil, which can sustain a good crop. In heavy floods,[65] which occur very rarely, it overflows its banks in the district of Mallani. The local name of this overflow is *rel*[66] and it is very fertile, as it is full of loamy soil. The entire region through which it flows, produce wheat and barley due to availability of water for irrigation. Its water is as a rule saline or brackish[67] but that of wells sunk at a distance of 20 or 30 yards from the banks of the river is usually sweet.[68] The inhabitants of all the villages situated in its neighbourhood depend on these wells for their drinking requirements.

The only important natural lake is the famous salt water lake of Sambhar.[69] Two other depressions of the same kind exist, one in the north of Jodhpur at Didwana,[70] and the other in the south at Pachbhadra.[71] There are a few *jhils* (marshes) in Marwar, notably one in the Sanchor district, which covers an area of 40 or 50 miles in the rainy season, and when dry can yield good crops of wheat and gram; its dry bed, being a good deposit of alluvium and moisture, yields good crops.

The physical characteristics of the region are somewhat complex and varied. The south-eastern boundary consists principally of metamorphic or transition rocks, rising precipitously from the plains of Jodhpur, and in some localities attaining an elevation of

[65] R.C. Temple (ed.), *The Travels of Peter Mundy in Europe and Asia (1608–1667)*, vol. II, Hakluyt Society, London, 1914, pp. 246–7.

[66] Nainsi, *Marwar-ra-Pargana-ri-Vigat*, vol. III, p. 136; Nainsi has used this term quite often to describe the nature of irrigation.

[67] Temple, *The Travels of Peter Mundy in Europe and Asia*, p. 248.

[68] Ibid.

[69] Das, *Vir Vinod*, vol. II, pp. 1250–1.

[70] Nainsi, *Marwar-ra-Pargana-ri-Vigat*, vol. I, p. 25.

[71] Ibid., p. 202.

3,000 ft. Passing from the Aravallis towards the west, the surface, even at the base of the mountain range, is found to be sandy but the understratum appears to be chiefly gneiss, hornblende, micro-slate and quartz. The aspect of the country therefore, as far as the Looni River is concerned, is that of a sandy plain, dotted with bold and picturesque conical hills. The most prominent of these formations are the Nudolai hills, the Punnagir hill near Jadhan, the Sojat hill; the hill near Pali, the Gundaj hill, the Sardars hill and the Jalor hill. Immediately around these hills the ground is hard and stony but gradually passes into sand, which becomes heavier as the eastern and northern districts are approached.[72]

The region that is spread between the Aravallis and Looni has a good forest cover. Trees of *jal* and *khejri* dominate the region. The dominance of these trees highlights the fact that they have adapted to these conditions. Similarly, the dominant livestock of the region—goats and camels—can eat these trees and thereby sustain themselves, even in the absence of fodder.

During rains this region gets dense vegetation. As discussed earlier, even light rains aid vegetation growth. Hence, during rains the growth is so thick that even sunlight does not pass through. Such a place was very good to arrange royal hunts, especially during the rains. There is no habitation in the nearby regions.[73] As soon as one comes out of the forest, one encounters *sathi kuwa*—a very deep well; its depth is equal to 60 men, which means a very low water table.[74] This description once again highlights the peculiarity of the region. Even though the water table is very low, yet, the forest of *jal* and *khejri* has survived.

After crossing the Looni, these conical hills are less numerous and sandstone appears. The country to the north of Jodhpur city is one vast sandy plain[75] called the *Thal*, only broken by sand-hills or *teebas*, which, commencing in this region stretch to Bikaner in the

[72] *The Rajputana Gazetteer*, vol. I, p. 20.

[73] बरसात छै, देस सुहामणो लागै छै । रावळजी हुकम करै तो थळ मांहे सिकार रमां । जठै घणा जाळ, घणा खेजड़, सूरज दीसै नहीं, अैड़ी झंगी तेथ गया । बसती कठै लाभै नहीं । इयै भांत सिकार रमै । Nainsi, *Munhot Nainsi ri Khyat*, vol. II, p. 285.

[74] Ibid., p. 286.

[75] Das, *Vir Vinod*, vol. II, p. 791.

north and to Jaisalmer and Sind to the west and south. Describing the region, Nainsi writes that the region is destitute. Further, he says that nothing much is grown here.[76]

The climate of this region is dry. This is due to the geographical location of the state, the geographical nature of the surface and the absence of a forest cover.[77] The Aravalli range separates it from the more fertile region of Mewar. The country is, therefore, beyond the range of the full force of the south-west monsoon which originates in the Indian Ocean, and is entirely removed from the influence of the south-east monsoon that originates in the Bay of Bengal. Also, the clouds from the south-west, before arriving at Marwar, float above the extensive arid and sandy tracts of northern Gujarat, Kutch, the Rann and the adjoining desert. This results in very little rainfall. All these conditions unite in precipitating extraordinary dryness in the region of Marwar. The sandy soil, the brackish water and the prevalence of the saline efflorescence known as *reh*, are the principal reasons for the scanty growth of wild jungle or crops on cultivated ground.

The last division of this arid region is Bikaner. It is situated north and east of Jaisalmer and Marwar and touches the more fertile regions of Punjab. The region contains no perennial rivers or streams. The southern and most of the north-eastern part of the state form part of the vast sandy tract known as *bagar*.[78] The north-west and part of the north lie within the Great Indian Desert. The north-east corner adjoining Punjab is the least unfertile section of the state.

> The northern pargana of the Bikaner state, lies along the bank of the Sotra or Hakra, in the dry bed of which wells are dug and the best water in the region obtained. The headquarters of the pargana Anupgarh, Sardargarh or Alwana, Suratgarh and Hanumangarh or Bhatner and Tibi are all on the banks of the ancient river. . . . The Naiwal is another Old Riverbed lying parallel to and few miles north of, the Sotra, but water never I believe flows in it. Similarly, the old bed of the Kagar, which crosses the district [Bhatner] and on which the fort . . . is situated, sometimes

[76] Nainsi, *Munhot Nainsi ri Khyat*, vol. III, pp. 5–6.

[77] Das, *Vir Vinod*, vol. II, pp. 790–3.

[78] Sharma, *Rajasthan Through the Ages*, p. 12.

contains water in rains, after which some wheat and barley is produced. . . . Tibi is situated on the east of Hanumangarh, it too is on the Sotra, and when the rains are good and the river flows and floods freely, good crops of wheat are subsequently raised.[79]

Bikaner suffers from extremes of heat and cold. During the hot season the heat is exceedingly great; heavy sandstorms are of frequent occurrence. In winters, the cold is generally very severe, trees and vegetation being injured by the frost. It is more due to its proximity to the Himalayas compared to other divisions of the region.

Water in Bikaner is found notwithstanding the slight apparent difference in the level of the country at varying depths, and is of very unequal quality.[80] Though the city wells are more than 300 ft deep, the water of most is of excellent quality. The situation of availability of water both in quality and quantity can be gauged from the following description of Bikampur region on the border of Bikaner and Jaisalmer: Nainsi informs us that there is a forty-men-deep well[81] in the fort of Bikampur. The water is salty. He also says that for 5 to 7 *kos*, either side, there is no water:

दोळो कोस 5 तथा 7 छै, नेड़ो कठै ही न छै।[82]

In this region water was so scarce that Nainsi has provided details of *talai*/tanks, wells in this region. He is also concerned about the duration of availability of water in a year.

ताळाई बिकूंपुररै नजीक

१ तिलांणी - कोस १, मास १ पांणी रहै ।

१ राणीवाळो - नोखव सेवड़ा बिचै, पांणी मास ४ ।

१ चंद्राव - भाटीरो, सेवड़ा थी कोस...., पांणी मास ४ ।

१ बह - सेवड़ा कनै, पांणी मास २ ।

१ बरजांग - जैतुंग सेवड़ा बीच कोस ३, पाणी मास ४ ।

१ गोपाळी - नींबली कनै, पांणी मास ४ ।

[79] Powlett, *Gazetteer of the Bikaner State*, pp. 121–4.

[80] Das, *Vir Vinod*, vol. II, p. 477.

[81] कोहर १ किडांणो प्रोळरी भीत हेठै छै, पांणी खारो छै, पुरसे ४०। Nainsi, *Munhot Nainsi ri Khyat*, vol. II, p. 113.

[82] Ibid.

१ हरख - जैसिंघ री सिरहड़, पाणी मास १० ।
१ गोधणली - सिरहड़ नजीक, पांणी मास ६, जूंनी छै ।
१ हरराज री लोहड़ी - सिरहड़ कनै, पांणी मास ४ ।
सिरहड़ तळाई ;१००द्ध कुथ ३ मीठा, पुरसै २० ।
जैतारी, पांणी मास ५ ।
भथरी, पांणी मास ४ ।
बावड़ी दळपतरी ।
तळाब रांणहळ, मास ८ पांणी। बेरा घणा ।
पूरादेरी, विकूंपुर वरसलपुर विचै कोस १२ ।
बीका सोळंकी रो, तळाबद्ध कोस ३ उत्तर नूं, पाणी मास ४ ।

तळाई बिकूंपुर रै देस -

१ अचलांणी - कोस १० रांणेरी कनै, पांणी मास ६ ।
१ नीबली - नींबा मुंहतारी, कोस १२, पांणी मास ४ ।
१ मांडळ - मांडा मुंहतारी, कोस ९, पांणी मास ४ ।
कांनड़ियारी-कांना सोढारी, कोस १०, रांगेरी कनै, पांणी मांस ।।
१ लूड़ी रांमसर बिकूंपुर था कोस...., पांणी मास २ रहै ।[83]

[List of various ponds near Bikampur, their distance from Bikampur and duration of availibility of water.
Tilani—1 *kos*, retains water for one month.
Rani Vano—between Nokh and Sewra, retains water for four months.
Chandrav—owned by the Bhatis, . . . *kos*, retains water for four months
Bah—near Sewra, retains water for two months.
Barjang—between Jaitung and Sewra, 3 *kos*, retains water for four months.
Gopali—near Nibnli, retains water for four months.
Harakh—near Sirhar, retains water for ten months.
Godhanli—near Sirhar, retains water for six months; it is an old one.
Har Raj ri Lohri—near Sirhar, retains water for ten months.
Pond of Sirhar—there are (100) wells at the depth of twenty hands, only three have sweet water.
Jaitari—retains water for five months.
Bhathari—retains water for four months.
Dalpat's step well
Ranamahal—retains water for eight months, sustains numerous wells.

[83] Nainsi, *Munhot Nainsi ri Khyat*, vol. II, pp. 134–7.

Pura Deri—between Bikampur and Barsalpur, 12 *kos*.
Bika Solanki's—3 *kos* north, retains water for ten months.
List of ponds around Bikampur
Aanchalni—10 *kos*, retains water for six months.
Nimbali—12 *kos*, retains water for four months.
Mandal—9 *kos*, retains water for four months.
Kandiyari—10 *kos*, retains water for two months.
Luri Ramnsar—. . . *kos*, retains water for two months.

There is another locality 16 *kos* from Bikampur towards Jaisalmer, known as Bap. The area surrounding this place is very fertile. It sustains a winter crop of wheat, which is grown on land inundated by rains. It is so fertile that if one sows a *man* of wheat, you can reap sixty times.

> बाप, बिकूंपुर सूं कोस १६, फळोधी, ठाकुराई री मंड बाप माथै छै । बाप घणा सैंवज गोहूं सारी सीव काठा नीपजै छै । मण १ गोहूं बायां मण ६० गोहूं हुबै छै । घणी ज्वार हुबै । सखरी साखर हुबै छै, ताहरां कण-नेपत गोहूं मण २००००० तथा ३००००० जाझेरा हुबै छै ।[84]

Bikaner was also known as *Jangal Desh*. As discussed earlier, from the viewpoint of vegetation the word *jangal* denotes a region where the sky is generally clear and water is scarce, and the land abounds in *sami* (*Prospis spicigera*), *karira* (*Capprisaphylla*), *peelu* (*Carya rborea*) and *kankandhu* (*Ciziphus jujuba*) trees. It was probably on account of being the rulers of the *jangal* area that the rulers of Bikaner were in a later period styled as *Jangal Desha Padshah*.[85]

There is a folklore in Rajasthan about why Rai Singh, the king of Bikaner, got his title, Bhurtio. It is said that a messenger from the Mughal court to Bikaner had to cross long stretches of desert, which were full of thorny bushes of *bharunt*. Rai Singh thus got the title Bhurtio.[86]

The existence of a fertile area in-between the arid region highlights not only the environmental diversity of the region

[84] Nainsi, *Munhot Nainsi ri Khyat*, vol. II, p. 137.
[85] Sharma, *Rajasthan Through the Ages*, p. 12.
[86] Nainsi, *Marwar-ra-Pargana-ri-Vigat*, vol. III, p. 123.

but also points out the possibilities of human settlement in an otherwise inhospitable arid region. It has to be kept in mind that though Bikaner is also part of the Great Indian Desert, the region sustains a fair amount of vegetation. The vegetation has adapted to the given environmental features and at a secondary level, animals and even human have adapted to the conditions. It is very clearly reflected in nature of the economy, where the pastoral component dominates. The details have been discussed in the subsequent chapters.

The only rocky hills in the state are at the borders of Jaipur and Jodhpur and even these are not more than 500 ft above the level of the plain. Towards the south-east from the city of Bikaner to the Jaisalmer border, the region is hard and stony. The greater part of the territory is plain undulating or interspersed with shifting sand-hills, whose slopes, lightly furrowed from the action of the wind, suggest the ribbed appearance of the seashore.

Bikaner has two small fresh water lakes. These are formed by the drainage of the rocky country south-west of Bikaner; they lie on the route from Bikaner to Jaisalmer and are famous as Kolayat and Gajner. Kolayat is an important pilgrimage place of the region. The salt lake of Chapar[87] in the Sujangarh area is 6 miles long by 2 miles wide but it is very shallow and almost dries up before the hot weather begins. There is another salt lake at Lunkaransar, [88] about 40 miles north-east of Bikaner.

Shekhawati region, adjoining Bikaner towards the east and south-east, receives very scanty rainfall. It is the last stretch of the little desert which extends from Jaisalmer. In Shekhawati, water lies at a great depth, averaging from 50 to 100 ft. It is brackish in many parts where the soil is impregnated with salt; but sweet water may be found throughout the east and south of the erstwhile state. There is normally one crop in a year, raised during the rainy season and ripening in October–November. This consists chiefly of *bajra*, *mung* and *moth*. The crop of the rainy season is the same as in the

[87] Nainsi, *Munhot Nainsi ri Khyat*, vol. II, p. 325; Chand, *Tawarikh Jaisalmer*, p. 477.

[88] Chand, *Tawarikh Jaisalmer*, p. 477.

north of Jaipur proper, but a little wheat and barley is grown in the cold season. Towards the south and east, as the soil becomes richer and firmer *jawar* and Indian corn with cotton and *til* are grown in the rains, while in winters wheat, barley, gram, sugar cane, opium, *dal*, tobacco and linseed are extensively raised. In the eastern districts, rice of a coarse quality is cultivated to a limited extent.

Moving towards the south-eastern division of Rajasthan, it is considerably smaller in extent than the others and consists of the higher and more fertile countries behind the Aravallis. In the central middle part of Rajasthan, the Aravallis lose the character of a distinct range; and north-eastward from this region they never altogether reunite, though their general direction is clearly indicated by successive hills and rocky eminencies as far as the group of hills near Khetri. Amid these disunited hills stands the town of Ajmer, on the highest level of an open tableland spreading on all sides. From Abu to Ajmer, the Aravallis offer a fairly clear line of demarcation between the sandy land of the north-west, and the more fertile region of the south-east; but beyond Ajmer the contrast is no longer so plainly marked.

In contrast to the sandy plains, which have more or less uniform features, this south-eastern division has a very diversified character. It contains extensive hill ranges, and long stretches of rocky mould, fertile valleys and woodland. These hill ranges are the source of the numerous rivers which traverse the region. In many parts there are wide vales, fertile tablelands and vast stretches of excellent soil.

As pointed out earlier, we shall be following the traditional political divisions for this region also. Mewar is one such division. Behind the loftiest and most clearly defined section of the Aravallis, which runs between Abu and Ajmer, lies Mewar. Towards the south-western corner of Mewar, the broken country behind the Aravallis is prolonged furthest into the interior.[89] Land in the region does not soften down into level tracts, but becomes entangled in a confused network of outlying hills and valleys, covered for the most part with thick jungle, which forms that very peculiar region

[89] Das, *Vir Vinod*, vol. I, pp. 102–3.

known as the hilly tract of Mewar. Major J. Rennell, while charting out the map of the region describes it as:

> The whole consists, generally, of high mountains divided by narrow valleys; or of plains, environs by mountains, accessible only by narrow passes and defiles: in effect, one of the strongest countries in the world; yet having a sufficient extent of arable land; of dimensions equal to the support of a numerous population; and blessed with a mild climate; . . .[90]

This tract has been described as an impregnable barrier.

Mewar is three-fifths plain and the remaining area is hilly terrain. A section of the Aravalli range of mountains extends over the south-western part of the state, from the city of Udaipur to the frontiers of Sirohi; where it stretches northwards through Kumalner towards Ajmer. Northward of Kumalner, this mountain tract is called Mhairwara and is full of ragged valleys and gorges.

Contemporary descriptions of Mewar at times exaggerate the goodness of the region and undermine the severity of the environmental features. To begin with, we shall analyse a description from a eulogy of *Med-Pat*, written by a Jain poet, Hem:

मेद पाट वर्णन प्रशस्ति

स्वस्ति श्री अवनि तिलक, मेद पाट विख्यात ।
देश सबे सिर सेहरौ, रोग नही तिलमात ।१।
देबां में सुरपति बड़ो, तारां में जिम चन्द ।
सरिता में गंगा बड़ी, गिर में मेरु गिरन्द ।२।
सबल देश में दीपतो, मोटो देश मेवाड़ ।
सरबर तरवर अति घणा, प्रोढ़ा अनल पहाड़ ।३।
पग-पग पाणी पंथ सिर, पग पग अम्बा रूंख ।
पग पग दीसै सेलढी, चखीयां जावै भूख ।४।
देश देशना मानवी, आबै जुगतै जात ।
चाबो चिहुँ ;दिसद्ध देशां मई, चतुर्भुज विख्यात ।५।
ढेबर सरवर जाणिये, कोशां बलि-चौबीस ।
राज समन्द निरख्यो अबस, बारां कोस बदीस ।८।
उदियासर अति दीपतो, पीछेलो परसिंध ।

[90] Major J. Rennell, *Memoir of a Map of Hindustan*, W. Bulmer & Co., London, 1793, pp. 230–1.

ठांम ठांम मन्दिर घणा, महारांणएकोध ।९।
सबै गढ़ां सिर सेहरी, गढ़ चित्रोड फबंत ।
गढ़ लंका संका धरी, जलधि जा जपंत ।१०।
कुम्भलमेर कट्टारगढ़, उँची अनढ़ पाहाड़ ।
सजल ठोढ बिखमी धरा, बाहुली झंगी झाड़ ।११।
देश अनोपम दीपतो, सोभा कही न जाय ।
एक मुखै करि किम कहुं, सहस मुखे कहिबाय ।१२।[91]

With the pronouncement of an auspicious word, I begin to describe the glory of Med-Pat–Mewar. It is the crown of all the lands, without a single fault. The way Indra is king of gods, the moon among stars, Ganga among rivers, Meru among mountains, similarly Mewar shines away among countries. It has got numerous ponds, tanks and a good forest cover, and along with it a beautiful and most ancient mountain series. The availability of water and trees in this land is immense. Sugar cane is produced in such a good amount that it is easily available and it is so sweet that one is sufficient to satisfy one's thirst. These are very famous. The ladies of the region are very beautiful. The lake of Debar is very large and Raj Samand is still larger. Udai Sar Lake is shining along with Pichhola Lake. The Maharana has constructed numerous temples. Among forts Chittor is supreme. As Lanka is situated beyond the sea, similarly to reach Chittor one has to cross the river. Kumbhalmer, Kattargarh are situated on the peaks of mountains and are secure due to growth of bushes with thorns all around them. It is difficult for me alone to describe the beauty of Mewar, it needs several people to describe the beauty of Mewar.

As it is a eulogy, we find a description dealing with the good things about Mewar, whereas Nainsi visualizes Mewar as a political unit and describes it accordingly. He has provided a detailed description of the mountains of this region. The description provides an insight into the contemporary social perceptions.

रूपजी-वासरोड़ देस रै फाळसे है । रूपजी सूं कोस ३ जीलवाळो दिखणनूं है । जीलवाळाथी कोस ३ रीछेर बीच अमजमाळरो बडो भाखर है । लाँबो कोस ५ है । उलै-कांनी कैलबो है। वाघोर रै आगै घाटो गांव है। तठा आगै घाटो गांव है । तठा आगे भोरड़ारो पहाड़ लांबो कोस ५ उतर-दिखण है । तठै भोरड़ नै मठावळा

[91] Agarchand Nahata (ed.), 'Desh Varnan', *Maru Bharti*, vol. II, no. 2, 1954, p. 49.

बीच समीचो गांव कूंभावतां सीसोदियारो उतन छै । उदैपुर से समीचो कोस १७, रूपजीथी कोस १२ छै । कूंभळमेर सूं कोस १० समीचो छै । तठा आगे मछावळो पहाड़ कौस ७ लांबो छै । गांव ९ मठावळा दोळा । मछावळा उपर पांणी घणो । झाड़ घणा । वेरणी नावै । तठा आगै वरवाडो । तठासूं वर नदी नीसरी छै । बनास नीसरी छै । तठा आगै घांसेररो मगरो कोस १ लांबी छै । तठा आगै पीडरझांपरो मगरो छै । घांसरे ने पीडरझांप बीच झांसनाळो कोनरो कोस २ छै । तठा आगे खमणरी मगरो छै । तठै लोहसीग गांव छै । तठै एक छोटी-सी नदी नीसरी छै । तठा आगै ईसवाळरो मगरो छै । गिरवारा भाखरांसूं जाय लागो छै । ईसवाळ उदैपुरसूं कोस ५ उतर पछिमनूं छै । जीलवाड़ा थी कोस ५ देसूरी । देसूरी थी कोस १ घांणेरो, कुभळमेर री तळेठी तठै । आगै कोस २ कुंभळमेर रो पहाड़ कोस १५ री गिरदवाय में छै । सादड़ी, राणपुर, सेवाडी तांई कुंभळमेर रो मगरो छै । सेवाड़ी कुंभळमेर सूं कोस ७ छै । तठा आगै राहगरो मगरो छै । निपट वडी ऐदी ठोड़ छै । पांणी पहाड़ मांहे निपट घणो छै ।[92]

Ropji-Vasroad is located at the entrance of Mewar. Jeelwaro is located 3 *kos* away from Ropji-Vasroad. Three *kos* east of Jeelwaro is Reechher. Reechher is located in the cut of the mountain Vaghor. There is Amjamal Mountain in-between Reechher and Jeelwaro. It is 5 *kos* long. Beyond it lies the Kailvo. There are villages in the valley of Vaghor. There is a 5 *kos*-long mountain Bhoraraโro lying north-south beyond Vaghor. All the villages between Bhorad and Machhawla in this region are part of the *watan* of Kumbhawat Sisodias. The region is 17 *kos* from Udaipur and 12 *kos* from Roopji. From Kumbhalmer it is 10 *kos*. There is the 7 *kos*–long Machhavalo Mountain beyond it. There are nine villages around Machhavalo. There is plenty of water available on this mountain. It is full of vegetation and its beauty can not be described. Beyond it there is Varvano mountain and river Var and Banas originate here. Beyond it lies the 1 *kos*-long hill of Ghanser. It is followed by the hill of Peeparjhhanp. Between these two, at an angle lies the Jhhansnalo. Beyond it lies the hill of Khaman. There is located a village known as Lohseeng. A small seasonal river originates from this place. Beyond it lies the hill of Eesval. The hill of Khaman culminates with the beginning of Girvara. Eesval is 5 *kos* north-west of Udaipur. Desoori is 5 *kos* from Jeelvaro. One *kos* from Desoori is Ghanerao in the valley of Kumbhalmer. The mountain of Kumbhalmer is stretched in 15 *kos*. The mountain of Kumbhalmer is extended till Sadri,

[92] Nainsi, *Munhot Nainsi ri Khyat*, vol. I, pp. 36–7.

Ranpur and Sewani. Sewani is 7 *kos* from Kumbhalmer. Beyond it lies the hill of Rahgaro. It is a very difficult terrain. The hill is full of water.

Along with the nature of vegetation in the region, Nainsi offers us specific information about the availability of water on the mountains. He has also tried to map the region by not only describing the respective locations of mountains but also the distances. Such a unique observation reflects upon the contemporary perception of the environment.

The general inclination of the region is from south-west to north-east, as indicated by the course of the principal river, the Banas,[93] and its numerous tributaries, mostly seasonal, flowing from the base of the Aravallis. Nainsi has traced the course of the river.

> जरगारा भाखरथी नीसरी । तिको जरगो उदैपुरुसूं कोस २९ छै । उठाथी रोहिड़ै गांव आवै । जको राजा हरचंदरो बसायो छै । उठाथी कोस २ गांव बरबाड़ो मेवाड़री तठै आवै । आगै कठाड़ गांव मदाररै गांव माछ में नै घांसाररै मगरे बीच नीसरै नै कांम-सकराही गांव वसै छै तठै आवै । उठाथी खभणोर आवै, उदैपुर थी कोस १२ । उठाथी कोठारिये आवै । तठा आगै गांव मोही तंवरॉंवले । तठा आगै जावद-नंदराय बीच नीसर नै गॉंव छै, उठै चोलेर रो पारसनाथ छै । उठा आगै पाड़लोळी जाजपुर रो गॉंव छै, तठै आय नै जाजपुर आवै । तठाथी सावड़ रै गॉंव देवळी आवै । आगै डावर तोडा रै गॉंव आवै, तठै खारी वधनोर वाळी भेळी हुई । आगै तोडा थी कोस गोकर्ण राह छै । बडो तीर्थ छै । मधुकीटभ तपस्या की छै । रावण तपस्या की छै तठै आवै । तठा आगै तोडारा गॉंव विसळपुर रावर आवै । तठै सोसादीये रायसिंघ मोहल कराया छै । तठा आगै बणहड़ै हुय टूंक आई । पछे मलीरणै रै गॉंव झूंपड़ाखैड़े सोहड़ भगवंतगढ सैसभारिजै मलीरणै रे वीछूंदे नै हुय जीरोत रो गॉंव हाडोती रो हुय नै आगै खंडरगढ चावळ भेळी हुई ।[94]

River Banas originates from the mountains of Jargara, which is 29 *kos* from Udaipur. It flows to village Rohire, founded by king Harchand. From there it goes to village Barvaro. It flows along the village Kathan and Machh. Then it intersects the hill of Ghansar and reaches the village of Kam-Sakrahi. Then it reaches the village of Khabhran, some 12 *kos*

[93] Nainsi, *Munhot Nainsi ri Khyat*, vol. I, p. 47.

[94] Ibid., p. 42.

from Udaipur. From there it reaches Kotherey and goes to village Mohi of Tunvaras. Then it bypasses the villages of Jadav–Nandrav and reaches the village of Choler of Parasnath. It reaches Jajpur after crossing village Parloli. It visits the village of Devli. Then it reaches village Dabar Toda where River Khari of Badhnor merges in it. Nearby is located the famous pilgrimage centre of Gokarn Mahadev. It is said that the infamous demons Madhukeetabh and Ravan meditated here. Afterwards the river crosses village Visalpur where Rai Singh Mohal, the Sisodia, constructed a palace. Then after crossing Banharen it reaches Tonk. In between it crosses village Jhunprakher and Beechhunde and moves on to Harauti by merging in the river Chambal.

The Banas originates from a mountain named Jargara, which is 29 *kos* from Udaipur and merges with Chambal in the Harauti region. Nainsi has traced the course of the river by identifying its course with major landmarks and villages. The mountains in-between are also mentioned in an attempt to locate it in his wider description of the natural features of the region. He has also described other rivers, which merge in-between.

To the south and west of Udaipur city, many streams emerge, which mostly flow through the Mahi Kantha southward into the Sabarmati River. Nainsi reports the following rivers in the region: Bhambhi,[95] Pagdhoi,[96] Juhi,[97] Var,[98] Berach,[99] etc. The physiographic features of the division provide ample scope for the construction of artificial tanks, etc. There are numerous natural and man-made lakes and tanks interspersed throughout the region.

Elaborating upon the pattern of habitation/villages in the region, Nainsi uses three categories: situated on mountains,[100] situated on a

[95] Ibid., p. 40.

[96] भैंसरोड़ हेठै चांमळ नदी वहै छै । तीन नदी भैंसरोड़रा कोट दोळी पिफरै छै । नदी तीन री विगत-१ चांवल । १ बांभणी । १ पगधेई ।, ibid., p. 40.

[97] वकरीया रो घाटो । जूही नदी उठै छै ।, ibid., p. 37.

[98] Ibid., p. 37.

[99] Ibid., p. 31.

[100] भाखर भांडेररो । कोस १० लाँबो, कोस २ पन्है । नाहेसर कोस १२ लो, पैनै कोस २ । तिण बीच जूडारो मुलक । गांव १ वांसै बीघा ५०० तथा ७०० खेती लायक । बीजी धरती भाखरो हठै । झाड़-आंबा, रायण पिण आंबली झाड़ घझा । घांन साळ, गोहूं, चिणा, मकी, उड़द घणा नीपजै । बालरण-काकड़ी घणी नीफ्रजे । दोबाँणरै नास-भाज बिखानूँ बडी ठोड़ इतरी ।, ibid., p. 37.

plateau[101] and others.[102] The region has an extensive forest cover.[103] The adjoining area of the erstwhile Dungarpur and Banswara state is still known as Vagadh and mentioned as such in the inscriptions of the locality. The name might have been derived, according to G.H. Ojha, from the word *bhagad*, which means forest, *jangal*.[104]

The Aravalli mountains of Udaipur are chiefly of granite, the valleys of variegated quartz. Slate is found in places. The metals known to exist in the low ground are lead and iron. Zinc mines are traced near Chittor.[105]

The south-western boundary of Rajasthan is formed by the state of Banswara. It is said to have taken its name from

> . . . the word means the country of bamboo (*Bans*). . . . The western portion of the state is comparatively open and well cultivated; but the rest of the country, especially in the south and east is covered with rugged hills, rocks, scrub jungle and woodland . . . more than half of the state is covered with jungle, the forest being dense in the northeast. . . . The climate is relaxing.[106]

To the east of Mewar is situated Harauti forming the southern boundary of Rajasthan. The Harauti region slopes gently northwards from the high tableland of Malwa, and is drained by the Chambal[107] with its tributaries, all flowing in a northerly or north-easterly direction. There are four major rivers in the region: Chambal, Sindh,[108] Par and Purar.[109] The Mokandarra range 1,200

[101] पठार रा गांव ८४ रतनपुरी चोरासी, चूंडावतांरी ठोड़ । गोहूं, चिणा नीपजै ९४ बेघमरौ बडो मुलक।, ibid., pp. 39–40.

[102] पीपळहड़ी सीरोड़रा मगरा, छ्पन चावड नै जवाछ, जावर वीच उदैपुरसूं कोस १७।, ibid., pp. 39–40.

[103] झाड़, पाहाड़, घणा छै । आंबा पूफलाद घणो ।, ibid.

[104] Sharma, *Rajasthan Through the Ages*, p. 17.

[105] जावररी खांण रूपारी रोज १ रु ४०० तथा ५०० आवै । जसद, रूपो नीसरे घोघंदो कोस ९ आथुणन । Further, he also informs us about the income from these mines. Nainsi, *Marwar-ra-Pargana-ri-Vigat*, vol. I, p. 32.

[106] *Imperial Gazetteer of India*, vol. VI, Clarendon Press, London, 1908, pp. 407–10.

[107] Das, *Vir Vinod*, vol. I, p. 101.

[108] नदी ४ हाडौती मांहै - १ चांवळ । १ सिंघ । १ पार । १ पुरड ।, Nainsi, *Marwar-ra-Pargana-ri-Vigat*, vol. I, p. 104.

[109] Ibid.

to 1,600 ft above sea level runs from the south-east to north-west, forming the southern border of the region. The hills are for the most part clothed with a thick jungle of striated trees and undergrowth.

The climate is moderately healthy. From November to February the climate is pleasant; in March it begins to get hot, and by the middle of June it is extremely sultry. The rains usually arrive in this area during the first half of July, and from then till the middle of October the climate is pleasant. Rainfall varies in the different parts of the region.[110] The northern part of the region is covered with the alluviam of the Chambal Valley. The central Bundi range divides this region into two parts. For many miles the precipitous scrap on the southern face of this range forms an almost impassable barrier between the plain country on either side.[111]

Describing the location of the city of Bundi, Nainsi says that the city is situated on the Aravalli range of mountains. It has good water supply. There is a tank, Sur Sagar, the overflow of which enables the inundated region to sustain a garden. The garden has mango, *champa* and other flowering trees. He also says that though there was no scarcity, the king's palace did not have sufficient water.[112]

> बूंदी सहर भाखर लगती बसै छै । रावळा-घर भांखरके आघोपरे छै । पिण मांहे पांणी मामूर नही । सहर आयो पीजै । सहर मांहै पाखती पांणी घणो । पीवै । बावे आंबा पूफलाद चंपा घणा ।

In the northern part of Bundi, the soil is for the most part hard, sustaining only a *kharif* crop, as agriculture is dependent on rains. The southern part of Bundi is rich in alluvial soils and it can produce any crop whereas towards Kota the land is fertile and well watered. The soils are divided locally into three classes: namely, *kali* a rich black loam containing much sand, *utar-mal*—a loam of a lighter colour but almost equally fertile; and *beri* a poor gravelly and sandy soil of a reddish colour, often mixed with *kankar.* On

[110] *Imperial Gazetteer of India*, vol. XV, Clarendon Press, London, 1908, pp. 410–11.

[111] *Imperial Gazetteer of India*, vol. IX, Clarendon Press, London, 1908, pp. 77–9.

[112] Nainsi, *Munhot Nainsi ri Khyat*, vol. I, p. 102.

the first two types of soils fine crops of wheat, gram can be grown without irrigation.[113]

Describing the *pargana* or the landscape of the Harauti region Nainsi informs us that there are 700 villages located in the plains and around 740 villages are located on the mountains.[114]

अगवारै गांव ७०० चौडै छै । पछबारै गांव ७४० भाखर झाड़ छै ।

As we move away from the Aravalli mountains the general character of the region of Amber is tolerably level and open, although its surface is crossed and diversified by groups and ranges of hills and by isolated peaks. The centre of the state is an elevated tableland of triangular form, from 1,400 to 1,600 ft above sea level, with a gradual slope to the south-east towards the Banas River. The eastern limit of this plateau is formed by ranges north and south and at places cut up by a broken chain of hills, an offshoot from the Aravallis.[115]

Towards the north-west the hills rise to a considerable height with a bold outline and form a natural boundary between the sandy desert tract of Shekhawati and Bikaner; to the south-west lies the more fertile soil of Jaipur proper.[116] On the east, beyond the range of hills near the city of Jaipur a gradual slope follows the valley of the Banganga River to the Bharatpur border, and the region becomes more open as it spreads out towards the alluvial flats of the Jamuna. In the extreme south of the state, the hills reappear. There is usually a fair amount of rainfall throughout the region. The Banas[117] receives most of the rainfall by means of several tributaries like the Banganga, Gambhiri,[118] Bandi, Monil, Dund, Khari, Kaotli,[119] etc., of which only one or two are perennial.

South of the dividing range between Shekhawati and Jaipur water is found beneath the surface at a depth varying from a

[113] *Imperial Gazetteer of India*, vol. XV, pp. 410–11; vol. IX, pp. 77–9.

[114] Nainsi, *Munhot Nainsi ri Khyat*, vol. I, p. 102.

[115] Das, *Vir Vinod*, vol. II, pp. 1246–7.

[116] Ibid.

[117] Nainsi, *Munhot Nainsi ri Khyat*, vol. I, p. 47.

[118] Das, *Vir Vinod*, vol. II, p. 1249.

[119] Ibid., p. 1250.

few feet in low lying land to 30–40 ft.[120] The soil of Jaipur in the immediate neighbourhood of the city and to the west and north also is generally sandy. In some places there are tracts of barren sand, frequently underlain by clay and stiff soil mixed with *kankar.*[121]

There are no forests of any significance in Jaipur, but the hills are more or less covered with *dhao* (*Anogeissus latifolius*) and other jungle trees of little or no value except for fuel. The *babool* (*Acacia arabica*) and *nim* (*Azadirchta indica*) may be considered as the principal trees of the region.[122]

In the northern part near the Khetri hills copper ore, cobalt and nickel are found along with the granite of the Aravallis. Cobalt is found within largess between the views of copper ore.[123] It is frequently used in Jaipur city for enamelling. Abundance of limestone is produced near the Karota and *kankar* (*Calcareous conglomerate*) is found almost everywhere, generally in flat beds.

The territory of Bharatpur forms the eastern boundary of Rajasthan. The general aspect is that of an immense alluvial plain, fairly well wooded and cultivated, with detached hills in the north, a hilly and broken district in the south, and low narrow ranges on parts of the western and north-eastern frontier. The principal rivers are Banganga, the Gambhir, Kakand and Ruparel; they usually cease to flow about two months after the rainy season is over. The soil of Bharatpur taken as a whole is probably superior to that of almost any other state in Rajputana; the territory further has the advantages of a good rainfall.[124]

Moving towards the north we encounter ridges of hills, for the most part parallel, and lying generally from north to south; these are a feature observable in the state of Alwar. The country to the north and east is generally open. To the north and west of Alwar, the soil is generally very light, except in parts on the western

[120] Boileau, *Personal Narrative*, p. 11.

[121] Das, *Vir Vinod*, vol. II, p. 1246.

[122] *Imperial Gazetteer of India*, vol. XIII, Clarendon Press, London, 1908, p. 391.

[123] Ibid., vol. VIII, Claerndon Press, London, 1908, p. 167.

[124] Ibid., pp. 72–4.

border, it does not form drift sand-heaps like those of Shekhawati. To the east there is much rich flooded land, but where water does not lie the soil is usually light. To the south the soil is generally pretty good.[125] From the point of view of geomorphology, the most conspicuous feature of the region is the Aravalli range of hills. Within the hilly ranges there are enclaves of fertile valleys and alluvial plains.

An interesting fact in relation to Rajasthan is the bearing that geographical features have on the name of a place. For example, *mer* in Rajasthani language connotes a hilly terrain and we see that places with an appellation *mer* like Ajmer, Barmer and Jaisalmer are towns that are located on a rocky surface. In Rajasthan, the importance of water in the location of settlements is emphasized with appellations like *sar*, *sir* and *tals*, which means water tank or pond in the local language. These appellations are more common in the north and north-western region where water was the most important factor behind any settlement, for example, Napasar, Bachasar (Bikaner), Roalsar, Sadosar (Churu), Bangasar, Rawatsar, Matsisar, Jabasir, Bhimsar (Jhun-Jhunu), Birmsir and Dhisar (Jaisalmer), Malsisar, Jabasir ka Tala and Buran ka Tala (Barmer), etc. It is pointed out that availability of water was the most important factor for the continuance of any settlement. Vigat is full of evidence that due to scarcity of water many villages were deserted.[126]

The delineation of a divergent landscape encapsulated in the broader political identity of Rajasthan and variations in the descriptions make it clear that to a great extent landsape is as much a social construction as it is an apparant physical feature. Though geographic features to a great extent define the broader contours of human intervention, we cannot negate the role of human agency. There can be no question of geographical determination; yet geography is the key, which can not be dispensed with.[127]

125 P.W. Powlett, *Gazetteer of Ulwar*, Sallantyne Press, London, 1874, p. 20.

126 Nainsi, *Marwar-ra-Pargana-ri-Vigat*, vol. I, provides details of villages which were deserted due to water scarcity.

127 Febvre, *A Geographical Introduction*, p. 49.

> We can say that "Man is a geographical agent". Everywhere he contributes his share towards investing the physiognomy of the Earth with those "changing expressions" which it is "the special charge" of geography to study. Through centuries and centuries, by his accumulated labour and the boldness and decision of his understandings, he appears to us as one of the most powerful agents in the modification of terrestrial surfaces. There is no power which he does not utilise and directs at will; there is no country, it has been said, which does not bear the marks of his intervention. He acts on the soil still more collectively—by means of all his groups, from the least to the greatest, from the family groups to the political ones. And this action of man and his environment is the "part" which man plays in geography.[128]

Therefore, the role of human ingenuity with specific reference to adaptations to given geographic features and the possibilities of the appropriation of given geographic characteristics is elaborated subsequently in the book.

[128] Ibid., pp. 63–4.

CHAPTER 2

Monsoon
Possibilities with Constraints

It is not an exaggeration to say that the monsoon personifies the weather of the Indian subcontinent. The emergence, growth and continuance of civilization in India cannot be imagined without the monsoon. Even today, in the modern industrial age any failure or even minor disruption in the routine of the monsoon plays havoc with agrarian production. Without state support it can lead to mass starvation. Indian agriculture for all practical purposes had been and is dependent upon the monsoon rains. Therefore, it is not surprising to see attempts at understanding the mechanism of the monsoon for prediction and if possible forecast. It is also not difficult to understand why prognostication of monsoon rains has gained immense significance in modern times. Moreover, it is interesting to note that attempts of prognostication of the monsoon have a long and complex history. Generally, philosophical thoughts about the environment have usually shaped the understanding of the monsoon and other climatic factors. A continuous inter-play between dominant philosophical explanations of nature and the realities of the monsoon rains can be traced till today. In ancient times, the monsoon was explained in terms of influences of visible factors like stars, constellations, the moon, the sun, etc. In the absence of tools to measure numerous features of the climate, attempts were made to explain climatic features in very general and vague terms. The uncertainties associated with the monsoon were usually explained as the influence of constellations or heavenly bodies. With the growth of modern methods of measurement and possibilities to examine innumerable physical factors in this computer age, it was expected that the riddle of the monsoon will be solved. However, improved capacity to document and corroborate local climatic features with regional and global variables has unfortunately enhanced uncertainty. That does not mean that modern science has failed in its endeavour, but it reminds us to recognize the complexity of climatic features. This chapter is an attempt to trace the way the monsoon has been understood over the centuries. It will also highlight the relationships between dominant philosophical traditions of the time and their influences on the understanding of the monsoon.

Despite the fact that two monsoon currents—the south-east and the south-west, cover the region, rains in the region have been erratic. Most of the rainfall is concentrated in the monsoon months from June to September. More importantly, in the analysis of the nature of the monsoon—of any particular year—for a region, one must be careful before generalization. Two documents of the same year for Jaipur state provide contrary images of the monsoon. An *arzdasht* by Rup Ram dated Ashadh Vadi 1, 1752 vs[1] informs us about drought, whereas for the same year Purohit Haras Ram writes in his *arzdasht* dated Bhadva Sudi 6 about good rainfall.[2]

It is not easy to describe the significance of the monsoon in the lives of common men as it is very hard to write about the obvious. Human dependence over the monsoon in India is so obvious that almost all aspects of social formations are directly or indirectly influenced by the monsoon. Nevertheless, to begin with, we shall delineate the popular perceptions of various natural phenomena. Folk culture is replete with praises and lamentations/complaints against the changing seasons. Folk traditions are at times satirical about seasons. They were aware of the inherent goodness of all the

[1] *Arzdasht*, Ashadh Vadi 1, 1752 vs, Historical Section, Jaipur Records, Rajasthan State Archives, Bikaner. Henceforth, HS, JR, RSAB.

[2] *Arzdasht*, Bhadva Sudi 6, 1752 vs, HS, JR, RSAB.

hardships. To begin with we shall analyse the response related to rains. As discussed earlier, due to the paucity of perennial rivers and the low water table, especially in the arid parts, the rains were the main source of potable water for humans and animals. Rains were eagerly awaited. The timely arrival of rains was very important. The following saying represents the desperation for rains. It is suggested that people are prepared to sacrifice their wealth to ensure rains:

> सौ सांढीया सौ करहलां, पूत निपूती होय ।
> मेवड़ला बूठा भला, होणी होय सो होय ।।[3]
>
> Hundred female camels can be sacrificed, sons may die, but we need rains.

and,

> मेह ने पावणां किताक दिनां रा ।
> राजा मान्या सो मानवी, मेयां मानी धरती ।[4]
>
> Rains occur for a very limited period and if the king is happy, everybody is happy; similarly land shall be happy only with the rains.

and,

> मोरिया तो मेहू मेहू करे,
> पर बरसणूं तो इन्दर कै हाथ है ।।[5]
>
> The peacock may cry the whole year but Indra—king of rains—manages rains.

and,

> मेहा तो त्यां बरसी, ज्यां राजी होसी राम ।।[6]
>
> Rains will take place only if god is pleased.

One of the sayings highlights the need of rains for agricultural production. The desperation of the common man is clearly visible

[3] Kanhiya Lal Sahal (ed.), 'Rajasthan ki Varsha Sambandhi Kahavaten', *Maru Bharati*, vol. IV, 1956, p. 18.

[4] Ibid., pp. 17–18.

[5] Ibid.

[6] Ibid.

in these sayings. It is further evident from the following saying where even the initial formation of clouds has been analysed in an attempt to predict the good and timely arrival of rains.

> जिण दिन होवै गरभड़ो, तिण थक्की छै मास ।
> उपर पनरा दीहड़ै, बरसै मेह सुगाज ।।[7]
>
> Rains occur due to formation of clouds which is identified as conception. The pregnancy lasts for six months and only then rains occur.

These sayings highlight the significance of rains in the life of the common man. Rains, though being a recurrent feature, depended to a great extent upon various other climatic factors, spread over the year. The anxious people of Rajasthan whose survival greatly depended upon the timely arrival of the monsoon attempted very carefully to relate the climate and rains. These were minutely observed.

CLASSICAL CONCEPTIONS OF MONSOON

The monsoon is a very complex climatic phenomenon. The word 'monsoon' owes its origin to an Arabic word meaning 'season'.[8] Though the term 'monsoon' for seasonal rains over the Indian subcontinent can be attributed to medieval times, as a climatic phenomenon it has a long history.

According to Sanskrit literature as well as to folk knowledge, the monsoon is indeed the result of a periodic reversal, but the realization of this reversal is with the sun and rains, not the winds. In the view of priests, astrologers and physicians—who were responsible for establishing the cycle of the festivals, the calendar of agricultural activities, the seasonal diets, and regimens—there exists a fundamental alteration between the Dry and the Wet principles. Monsoon comes after the summer solstice, when Agni (the sun) reverses its course and the earth comes under the rule of Soma (the moon), the dispenser of rain.[9]

[7] Ibid., p. 13.

[8] Das, *Indian Monsoon*, p. 1.

[9] Zimmerman, 'Monsoon in Traditional Culture,' p. 51.

Such an explanation is in consonance with the contemporary understanding of the cosmos. It is suggested that Indian cosmology was excessively obsessed with the celestial bodies and its influences on human existence. The sun, moon and seasons showed a remarkable cyclic pattern. Similar cyclic patterns were also visible in the wider space, so it was but natural for them to explain changes in climate only in terms of noticeable objects/features. The visible cosmos and landscape shaped their vision of climate and changes associated with it. Such a vision of the universe also corresponded to the agrarian cycle. The heat of the sun during the summers inhibited possibilities of agrarian production, whereas rains made agriculture possible.

Understanding of the monsoon in ancient India to a large extent corresponded to the prevalent vision of cosmology. Meteorological observations were also coloured by the people's vision of the cosmos where the sun, stars and moon provided the tools to develop their almanac—the calendar. The Indian almanac was a manifestation of immaculate mathematics to calculate the relative position of numerous stars with respect to the sun and the moon.

> The native calendar ties the seasons and climates to one of the most important series of astrological symbols, the 27 *naksatras*. The *naksatras* —constellations or asterisms—correspond to the lunar mansions, and to fixed period of the sun's stations on the lunar zodiac. . . . The *naksatras* define the lunar zodiac, that is, the cycle of moon's stations. The positions of the sun, in its course over the ecliptic, are also defined to these asterisms resulting in the 27 divisions of the year.[10]

Given the regular cyclic nature of celestial bodies, and to a great extent even seasons, a correlation was in plain sight, and it justifies their reliance on the almanac for the description of the timings of the monsoon in India. As closer examination of few popular sayings from the *Madras Glossary* suggests, farmers believed that if in Karttika there is rains then one can start ploughing. A dry Mrgasiras symbolized a paucity of rains until the middle of August. The most sought after *naksatras* has been Ardra as it

[10] Ibid., p. 69.

marks the beginning of full-fledged monsoon in norther India, but it was considered inauspicious for either sowing or transplanting. Peasants were advised in the proverbs that they can plan second sowing of paddy during Punarvasu, if the first has failed. Similarly, it was suggested that if there is thunder and lightning in Ardra, there will be no rain for sixty days. However, if it rains in Ardra, it will continue for a long time.[11]

Commenting on the significance of these, he terms them as, 'The figures dogmatically set forth ("for sixty days") sum up an immemorial tradition of meteorological observation.' Further, by examining the close relationships between the almanac and meteorological observation Zimmerman noticed that the classical astrological texts establish a scheme for rain forecasting in which the *naksatras* do not represent the fixed periods of the sun's stations on the lunar zodiac, but the lunar mansions proper. It was believed that monsoon has a six-month long gestatation period, therefore one can forecast more than six months in advance by examining the nature of clouds under a particular station of moon.[12]

Such an explanation was in tune with the vision of the cosmos as a human being where the cycle of the monsoon was equated with the birth of a child. It was firmly believed that the life of humans will be governed by the place and time of birth in a given constellation. The future of the child was predicted in terms of a horoscope. Similarly, rains during the monsoon were predicted according to the horoscope developed in terms of time of conception. It also has resonance with the agricultural production where the earth was treated as a womb.

The monsoons in India are the mainstay of agrarian production and most of the times the fertility of the soil is inhibited by the absence of rains. Agricultural yield was believed to be grown from the womb of the earth and that could take place only when rain showered its benevolent seeds. Agrarian production was depicted as organic growth with a certain amount of gestation period once the seeds of rains (water) had fertilized the earth. Corresponding

[11] Ibid., p. 70.
[12] Ibid.

characteristics were attributed to the clouds as well. Zimmerman suggests that:

> The symptoms of pregnancy must be detected in the period starting from the day when the moon reaches the asterism Purvasadha during the bright fortnight of the month of Margsirsa. The foetus formed during the moon's stay in a particular asterism will deliver rain 195 days after, when the moon passes through the same asterism according to the laws of its revolution.[13]

It is further evident from the following saying where even the initial formation of clouds has been analysed in an attempt to predict the good and timely arrival of rains: 'Rains occur due to formation of clouds which is identified as conception. The pregnancy last for six months and only then rains occur.'[14]

Monsoon or seasonal rains was traditionally explained in terms of interplay between celestial bodies including the sun, and the moon. Therefore it was necessary to understand and map the movements of celestial bodies. Astronomical observations and calculation of movement was seen as one of the tools to understand the climate or seasons. India has a long tradition of astronomers who made important contributions. 'During the fifth century we have the outstanding Indian mathematician-astronomers Bhaskara and Aryabhatta. Bhaskara also developed an instrument, which was like a simple astrolabe. During the sixth century, India had other outstanding astronomers in Varahmihira and Brahmagupta.' Therefore, astronomical observations had achieved very high standards.[15]

However, developments in meteorology were more pronounced in Greek Philosophy. The way visible features of the landscape and the universe influence the conception of climatic features can be better comprehended, if we examine the Greek comprehension of climate.[16]

13 Ibid., p. 70.

14 Sahal, 'Rajasthan ki Varsha Sambandhi Kahavaten', p. 13.

15 A. Rahman, *Maharaja Sawai Jai Singh II and Indian Renaissance*, Navrang, New Delhi, 1987, p. 15.

16 Liba Taub, *Ancient Meteorology*, Routledge, London and New York, 2003.

Aristotle divided the universe into two major regions: the celestial region beyond the orbit of the moon, and the terrestrial or sub-lunar region—sphere of the moon's orbit about the earth. Thus, he made a precise distinction between astronomy and his new subject, meteorology. The former was restricted to the celestial region (including the orbit of the moon); the latter was restricted to phenomena of the terrestrial region.[17]

The diurnal effect of the moon on the sea to cause tides must have been visible to any coastal society and the powerful role of the moon influenced the Greek vision of the universe. The universe was seen in two separate spheres where one is attributed with the power to influence the climate and the other is relegated as passive. This demarcation helped Greeks to comprehend the nature of the monsoon in terms of the reversal of winds. Arrian in his book *Anabasis Alexandri*, while describing the preparations for Nearchus's voyages almost identifies the monsoon in terms of the *reversal* of winds near the mouth of the Indus in autumn.

The season of the year was then unfit for voyaging; for the periodical winds prevailed, which at that season do not blow there from the north, as with us, but from the Great Sea, in the direction of the south wind. Moreover, it was reported that there the sea was fit for navigation after the beginning of winter, that is, from the setting of the Pleiades (a star cluster which set at sunset on about 4 November at the latitude of the Indus delta in Alexander's time) until the winter solstice; for at that season mild breeze usually blow from the land, drenched as it has been with great rains, and these winds are convenient on a coasting voyage both for oars and sails.[18]

By closely examining the writings of Arrian, Bruce A. Wareen has pointed out that perhaps Arrian was suggesting a plausible link between rains (monsoon) in India and rains in Egypt. Arrian writes:

[17] H.H. Frisinger, *The History of Meteorology: To 1800*, Science History Publications, New York, 1977, p. 15.

[18] As quoted in B.A. Warren, 'Ancient and Medieval Records of the Monsoon Winds and Currents of the Indian Ocean', in Jay S. Fein and Pamela L. Stephens (eds.), *Monsoon*, John Wiley & Sons, New York, Chichester, Brisbane, Toronto, Singapore, 1987, p. 141.

India is visited by rain in the summer, especially the mountains . . . and from these the rivers flow swollen and muddy. In the summer also the plains of India are visited by rain, so that a great part of them are covered with pools; and Alexander's army had to avoid the river Acesines [Chenab river, a tributary of Indus] in the middle of the summer, because the water overflowed the plains. Wherefore from this it is possible to conjecture the cause of the similar condition of the Nile, because it is probable that the mountains of Ethiopia are visited by rain in the summer, and the Nile being filled from them overflows its banks into the Egyptian country.[19]

Any further conjunction on the origin of the monsoon was constrained by Aristotelian thinking:

who claimed that Water could "flow" only in its "natural direction", which like that of Earth, was downward (the natural direction for his other two primary elements, Air and Fire, was upward): thus the ocean, occupying the lowest places, could only "swing to and fro". Hence the monsoon currents are conceived here essentially as an annual tide. Nevertheless, the description is roughly accurate for the north Indian Ocean with respect to both direction of flow and season.[20]

The theories proposed, with the possible exception of cloud movements, were not verified by measurements of the phenomena they concerned. They could be accepted or rejected only by being related speculatively to more general theories of nature as the theory of the four elements.

The four elements—fire, air, water and earth—were not the (somewhat) irreducible substances. They were qualities, or forms of existence, that changed into one another or mixed to create new qualities. The scheme proved richly nuanced and unfortunately perfectly suited for speculation about the sky. Sunshine triggered a series of transformations between elements that brought about weather. Ironically, this concept was close to our understanding of the weather today, for we know that the sun drives the cycling of moisture and heat and air around globe. The varying angles of sunlight force the air to distribute unequal solar heating. The unequal heating of different ground cover—rock, trees, water, and sand—also creates wind.[21]

19 As quoted in ibid., p. 142.

20 As quoted in ibid., p. 149.

21 Frisinger, *The History of Meteorology*, p. 10.

Aristotle's world worked differently, even if he took similar factors into account. For instance,

> the heat of the sun's rays incited the Earth to emit a rising exhalation, a hot and dry substance prefiguring fire. Similarly, the oceans—the layer of the world overlaying the solid Earth—emitted hot, moist vapours. These exhalations and vapours were lighter than their parent elements and sought their proper, higher level in universe.[22]

Prognostication from the observation of certain astronomical and atmospheric phenomena also appears in Theophrastus's work. Many shooting stars are considered a sign of rain or wind. Observing the moon is also important, that is, 'If the moon looks fiery, it indicates breezy weather for that month, if dusky, wet weather.' It is interesting to observe that nearly all of the still popular empirical rules for weather forecast (reddish sky at sunrise foretells rain) come from this short work. Theophrastus made no attempt to explain the different atmospheric phenomena, but referred all such consideration to the Aristotelian method. While Aristotle's work was, thus, largely theoretical, Theophrastus's short treatise was completely practical. Theophrastus of Eresos wrote *De Signis Tempestatum* (on Weather Signs) and a treatise, *De Ventis* on winds. The practice of weather prognostication by some eighty different signs of rains, forty-five of winds, fifty of storms, twenty-four of fair weather, and seven signs of weather for periods of a year or less. . . . It is the oldest collection of weather signs to have survived, and most of the later collections were based on it.[23]

TRADITIONAL NOTIONS OF MONSOON

The uncertainty associated with seasons, winds, rains, etc., have caused immense anxiety among the peasantry since ancient times. In the absence of scientific instruments to measure or record the meteorological features, it was but natural to rely on other broad factors visible in nature. Naturally this led to the growth of a series of conjunctures based on the different permutations

[22] J. Rosenfeld, *Eye of the Storm: Inside the World's Deadliest Hurricanes, Tornadoes and Blizzards*, Plenum Trade, New York and London, 1999, p. 28.

[23] Frisinger, *The History of Meteorology*, pp. 25–6.

and combinations of those visible factors. Such observations were converted into popular sayings. There are innumerable references, yet before any discussion on them, it is important to keep in mind that such beliefs may not be scientifically provable. The observations made by common man were actually based on visible features or the features he was aware of. There might have been several other factors of which society was not aware of or which were not visible to them. Thus, we can say that these popular sayings were based on observations and were a part of their experience. Although all of these sayings may sound superstitious, they did represent a complex web of meanings and values, which the rural folk attached to different phenomena of nature. Their frequent reference in the literature indicates not merely their popularity but also the significance attached to them in the popular domain in terms of contemporary comprehensions of weather and climate.

We have numerous popular sayings representing a combination of factors like wind patterns, sky, clouds, dust storms, etc., to conjuncture about the future climate. The variables considered were further expanded to the immediate climatic factors. It was believed that if wind is blowing from north-west, then there is a definite possibility that it will rain. Similarly, it was suggested that if the sky is blue during monsoon season, it will rain. Relying on the nature and timing of lightning it was believed that if there was no lightning during the month of Chaitra, then rains shall start in the month of Baisakh. Further, it was postulated that if a rainbow is visible in the morning and sunrays are red in the evening, it shall lead to floods.[24]

Predictions were carried out often on the movements of celestial bodies also. These include not only the sun and the moon but also the other planets and stars:

सोमां सुकरां सुरगुरां, जे चन्दो उगन्त ।
डंक कहे है भड्डळी, जळ थल एक करन्त ।।
सावण तो सूतो भलो, उभी भलो असाढ ।

[24] Sahal, 'Rajasthan ki Varsha Sambandhi Kahavaten', p. 18.

मंगल रथ आगे हुबे, लारे हुबे जो भान ।।
आरभिया यूं ही रहै, ठाली रैबै निवाण ।।
सूरज कुंड अर चांद जलेरी ।
टूटा टीबा भरगी डैरी ।।
आदरा भरै खाबड़ा, पुनरबसु भरै तळाब ।
न बरस्यो पुषै तो, बरसही घणा दुखै ।।[25]

If in the month of Ashadh, Moon appears either on a Monday or on a Thursday or Friday, then it will rain. In the month of Sawan, Moon is not bright on the second day and in the month of Ashadh if it is bright, then it is a good sign. However, if the Moon rises before the Sun in these months then there shall be drought. If there is a ring around the Sun and a Jaleri around the Moon then rains shall be so heavy that sand dunes will flow down. Similarly, if rains take place in the constellation of Adra then it shall be heavy, and if it begins in the constellation of Pushya, it shall not be a good year.

The anxiousness to know whether there would be good rain in the forthcoming season was a very important concern, as any failure of the monsoon would have led to drought. The concerns for timely and good rains are visible in the prevalent folklores. An attempt was made by people to predict the rains for the forthcoming season on the basis of different natural features spread over the months of the year:

काती सुद पूनो दिवस, जे क्रितिका रुख हुन्त ।
जे बादळ बीजू खबै, चार सास बरसन्त ।।
मंगसर तणी'ज अस्टमी, बादळ बीजां होय ।
सावण बरसै भड्डली, साख सवाई जोय ।।
पोस अंधरी दस्समी, चमकै बादळ बीज ।
तो भर बरसै भादवो, सायध खेलै तीज ।।
माह'ज पड़वा उजळी, बादल वाव'ज होय।
तेल, धीव अर दूध सब, दिन-नि सू'ध जोय ।।
फागण बद दुनिया दिवस, बादळ होय'स बीज ।
बरसै सावण भादबो, चंगी हीबै तीज ।।
नव दिन कहिजै नौरता, सुकल चैत कै मास ।

[25] Ibid., pp. 10–11.

जळ बूठै बिजळी हुवै, जाणो गरभ बिनास ।।
बद बसाख अमावसी, रेवति होय सुगाळ।
मध्यम होवै अस्विनी, भरणी करै दुकाळ ।।
जेठ बदी दसमी दिवस, जे सनि वासर होय ।
पाणी होय न धरण में, बिरळा जीवै कोय ।।
पैली पड़वा गाजै, तो दिन बहोत्तर बाजै ।
सावण पैली पंचमी, जो धड़ूकै मेव ।।
च्यार सास बरसै सहो, सत भाखै सहदेव ।
भाद्रव छ्ट छूट्यो नहीं, बिजळी रो झणकार ।।
तू पिय! जायै माळवै, हूं जाउँफ मौसाळ ।
ध्रुर आसोज अमावसां, जै आवै सनिवार ।।
समयौ होसी करवरौ, पिंडत कहै विचार ।
मूल्या फिरै गँवार, कानी भाळें मेहड़ा ।।
माघ मसक्कां-जेठ सो, सावण ठंडी वाब ।
भीम कहै सुख भड्डली, नहीं बरसण रो दाब ।।[26]

If the constellation of Kritika is in the *sudi* fortnight (bright fornight) of the month of Karttik and lightning takes place, it shall continuously rain for four months. If on the *vadi* (black fortnight) eighth day of Margshish there is clouds and lightning, it shall rain in Sawan and harvest shall be one quarter more. If on the *vadi* tenth day of Paus there is lightning in the clouds then there shall be good rains in Bhadra. If in the *sudi* fortnight of the month of *pratipada* it is windy and cloudy, then monsoon shall be a failure. If on the *vadi* second day of Phalgun, there are clouds and lightning then it shall rain in the months of Sawan and Bhado. If there are rains during the Navratri of the bright fortnight of Chait, then it shall be a drought year. In the month of Baisakh, on the day of the new moon in *vadi* and if it is in the constellation of Revati, it shall be a good year, if it is Ashwani, it is an average year and if it is Bharni, then the year shall be drought. If on the *vadi* tenth day of the month of Jeth, it is Saturday then there shall be no rains and monsoon shall fail. If on the *vadi pratipada* of the month of Ashadh, clouds are thundering then it shall rain for seventy-two days continuously. If on the *vadi* fifth day of Sawan clouds are thundering then it shall rain for four months. If on the sixth day of Bhadrapada, lightning does not take place, then people have to migrate toward Malwa. If there is New Moon on the *vadi* of Asoj, then rains shall

[26] Ibid., pp. 12–13.

be average. Only fools expect rains in the month of Karttik. If days are hot in the month of Magh, cold in Jeth and cold wind blows in Sawan, then it shall not be a good year for rains.[27]

It is significant that the linkages between various physical features of the environment and rains included movements of celestial objects too. As most of the celestial bodies follow a cyclical pattern spread over the whole year, numerous linkages have been observed. It will be interesting to examine in detail the role of the *panchang*[28] in these linkages. The significance of the *panchang* can be gauged by the fact that celebration of numerous festivals and *vrats* (days for fasting and worship) on specified days (the calculation for these is based on the movements of celestial bodies) was/is based on the reasonable understanding and greater role of the almanacs in social life.

These sayings clearly bring out the linkages among various environmental features which man had developed to predict the nature of rains in the forthcoming season in India. Simultaneously, it is important to emphasize that these correlations between the natural environment, social practices and the degree of probable rainfall were rooted in the actual physical context of the region concerned. The observations were even based on the activities of animals.

These observations can be classified according to the variables under consideration.

अत पित वाळो आदमी, सोवे निद्रा घोर ।
अणपढ़िया आतम थकी, कहै मेघ अति जोर ।।
वात पित्त युत देह ज्यां, होय रहे धम धूम ।
अणभणिया आगम कथे, रहे मेह की धूम ।।[29]

[27] Ibid.

[28] An almanac based on a lunar calendar. It contains information on movement and positions of various planets and other celestial bodies, the configuration of which at any given point of time is supposed to effect temporal events and the almanac presents all such information in a tabular form.

[29] Ibid.

It is said that if a person with a tendency of *pitt* sleeps well and a person with a tendency of *vat* suffers from a headache due to the heat then rains shall be very heavy.

Similarly,

कुन्दन जमै नं जड़ाव प, जमसलायन कीट ।
कहे जढ़िया सुणजो जगत, उड़े मेह की रीठ ।।
धेब्यां धेखो मिट गयो, मन में हुबो हुलास ।
देख सूदणी बजबजी, मेह आबण की आस ।।
कोरा कपड़ा सूदणी, जद अत गर्मी होय ।
सूछ्म कीड़ा सूक्षणी, मेहा मुक्ता जोय ।।
बणकर केरी पांजनी, सूखे नहीं सताब ।
आबादानी मेह की, लाल रंग वहै आभ ।।
देख खुरड़ कहे ढेढ की, कथा टूटे नेह ।
ल्हेई चढ़े न चामड़े, मुक्ता बरसे मेह ।।
ढोल दमामा दुड़बड़ी, बोरे सादर बाज ।
कहे डोम दिन तीन में, इन्द्र करै आवाज ।।
बिगड़े धिरत बिलोवणो, नारी होय उदास ।
जद असवारी मेह की, रहें छास की छास ।।[30]

When a gem cannot be fixed on ornaments, insects infest on wood sticks, it is going to rain heavily. When clothes do not take starch, wet cloths get infected with tiny insects then it is going to rain heavily. When woven cloth does not retain *pan*, leather does not bind with glue then it is going to rain heavily. Similarly, if musical instruments like the drum, *tasha*, etc., do not give proper sound then it is going to rain heavily.

These sayings clearly bring out the linkages among various environmental features which man had developed to predict the nature of rains in the forthcoming season.

The observations were even based on the activities of animals:

आम सूजै सांढणी, दौड़ै थळां अपार ।
पग पटकै वैसे नहीं, जद मेह आवणहार ।।
सावण काछा झाग सुण, गाडर हंदा हंत ।
दौड़े सनमुख पवन दिस, जळ थळ ठेल भरन्त ।।

[30] Ibid.

मांडे राड सांप री मासी ।
तो जाणी चोकस मेह आसी ।।[31]

It means that if a female camel runs around, throws her legs around, a sheep spits foam from her mouth and runs in the direction of the wind then rains shall be heavy.

Similarly, the activities of birds were also closely observed:

चिड़ी'ज नहावे धूळ में, मेहा आवणहार ।
जळ में नहाबै चड़कली, मेह विदा तिण बार ।।
बग पंखां पैाय, उझकि चोंच पवनां भखे ।।
तीतर गूंगा थाय, इन्द्र धड़ूके माघजी ।।
टीळै मिलकी कांबळी, आय थळां बैठन्त ।
दिन चौथे कै पाँचवैं जल थळ ठेळ भरन्त ।।
पपैयो पिउ पिउ करै, मोरां धणी अजग्ग ।
छत्रा करै मोरयो सिरै, नदियां बहै अथग्ग ।।
अत तरणावै तीतरी, लक्खारी कुरळेह ।
सारस गिरी-शृंगन भ्रमें, जद अत जोरे मेह ।।[32]

When birds play in the dust, it shall rain. When small birds play in water, the rains are over. When the *bagula* stretches his wings and breathes through his beak, *titar* is silent, *cheelen* sit on land, *papiha* and peacocks start crying and dancing, a *saras* starts flying high, etc., then it is time for heavy rains.

The activities of insects were also closely observed:

सांप, गोयरा, डैडरा, कीड़ी, मकोड़ी जाण ।
दर छाड़े बाहर भमे, नहीं मेह की हाण ।।
गिरगिट रंग, बिरंग हो, मक्खी चटके देह ।
माकड़िया चह-चह करे, जद अत जोरे मेह ।।
कीड़ी मुख में अंड ले, दर तज भूमि भमन्त ।
बरखा तु विशेष यो, जळ-थळ ठेळ भरन्त ।।[33]

When snakes, frogs, ants, spiders, etc., come out of their dwelling in the land, a chameleon changes colour quickly, flies stick to the skin of humans and ants carry their eggs, then it is going to rain heavily.

[31] Ibid.
[32] Ibid., p. 10.
[33] Ibid.

We have already pointed out the panorama of interrelated environmental features to predict the rains. A singular effort can be noticed in these sayings. It also signifies that common man was closer to the natural environment and was not confined only to the social environment. The activities of various animals, birds and insects were also relied upon to establish a pattern of relationship and inter-linkages with rain.

The variables considered were further expanded to the immediate climatic factors. Such observations of immediate climatic factors were an important part of the sayings:

बिजनस पवन सूरियो बाजे ।
घड़ी पलक महिं मेह गाजे ।।
पवन गिरी छुटै परवाई ।
धर गिर छोलां इन्द्र धपाइ ।।
सवार रो गाजियो, ऐळो नहीं जाय ।
प्रात:काल बादल का गरजना वृथ नहीं जाता ।।
बादळ रहे रात रा बासी ।
तो जाणों चोकस मेह आसी ।।
सुकरवार री चादरी, रही सनीचर छाय ।
ढंक कहै हे भड्डुळी, बरस्याँ बिना न जाय ।।
अम्मर राच्यो, मे माच्यो ।
तारा अत तग-तग करे, अम्बर नीला हून्त ।।
पढ़े परळ पाणी तणी, जद संजया पूलन्त ।
अस्भर पीळो, मे सीळो ।।
चैत महीने बीच लुकोवे ।
धुर बैशाखां केसू धेवे ।।
उगंते रो माछ्ली, आंधवते रो मोरव ।।
डंक कहे है भड्डुली, नदियां चढ़सी गोरव ।।
आँधी साथै मेह आया ही करै ।
आँधी रांड मेहां रो पाळी रैवै ।।[34]

If the wind is blowing from the north-west, then it will rain. If the wind blows from the east rains shall be heavy. If clouds thunder in the morning then it will rain. If clouds of the previous night continue, then it will rain. If clouds appear on Friday and continue even on Saturday, then it will

[34] Ibid.

certainly rain. If the sky is red it will rain. During the monsoon season if the sky is blue it will rain. If the sky is yellow, rains will mellow down. If there was no lightning during the month of Chaitra, then rains shall start in the month of Baisakh. If a rainbow is visible in the morning and sunrays are red in the evening, it shall lead to floods. Rains accompany dust storms. Dust storms subside only with the visit of rains.

Many sayings establish a correlation between the growth of plants and climatic phenomenon. As stated earlier, it is said that 'if the fruits of the *neem* tree dry on the tree itself and do not fall to the ground it is going to be a famine year'.

नीम्बोली सूखे नीम पर, पड़े न नीचे आय ।
अन निपजै नहिं एक गण, काळ पड़ैलो आय ।।[35]

It was a common practice to relate the abundance or failure of the growth of various plants, flowers, etc., to the nature/extent of the monsoon.

वन-बेरी, फूलै फलै, यूं खेजड़ ढहग्गट्ट ।
नहिं अंकुरै-बड़ जटन, वहै दुर्भिच्छन हग्गट्ट ।।
अंकुर, फूटे बड़ जटा, बैशाख जेठ के मांय ।
बैत डौढ परमाण तो, समयो आछो थाय ।।
जो लम्बी आ होवे नही के बेगी ही बधजाय ।
अल्प मेह बिरवा घणों, के देरी करके आय ।।
बन बेरी अरु खेजड़ी, अर्धपात झड़जाय ।
अर्धपात साबत रयां, करसण समो कहाय ।।[36]

If the small trees of *ber* and *chhonk* grow densely in the month of Asad then it shall be a monsoon-deficient year. However, if half the leaves fell down and the rest remained on the tree, then agricultural production for the year shall be on an average. Similarly, if the *bargad* tree does not flower, it shall be a bad year.

These sayings once again reflect the close human-environment interaction. An attempt was made to relate various natural phenomena like the changes in vegetation with the advent of

[35] Jai Shankar Dev Shankar Sharma, 'Prakriti se Varsha Gyan ki Banagi', *Rajasthan Bharti*, vol. VI, nos. 3–4, p. 59.

[36] Ibid., pp. 60–2.

rains in the next season. A closer look at the preceding sayings clearly brings out the fact that man was well aware of the natural vegetation around him. Such an outlook extended to the wild vegetation too.

Although all of these sayings may sound superstitious, they did represent a complex web of meanings and values, which the rural folk attached to different phenomena of nature. Their frequent reference in the literature indicates not merely their popularity but also the significance attached to them in the popular domain in terms of their own perceptions of weather and climate.

These examples clearly explain that man was much closer to the natural physical world. He had created a world of beliefs based on observation of the various cycles of nature. The popular sayings were a medium to express his understanding of the complex interdependent relationships of numerous physical factors present in his immediate and distant environment. Similar correlations were also seen in Europe. It was believed that animals and birds are more close to nature and therefore are more sensitive to understand the climatic patterns and changes associated with it.

> Unusual actions of animals and birds have for centuries been observed as indications of future weather. The still popular empirical rules, based on the actions of animals and birds, are enumerated in this treatise. Such rules as 'It is a sign of storm or rain when the ox licks his fore-hoof,' 'A dog rolling on the ground is a sign of violent storm,' and 'It indicated an early winter when the breeding season of sheep begins early' are quite familiar to many a present day farmer.[37]

INVESTIGATING MONSOON

Further development in the prognostication of the monsoon could take place only with developments in the field of meteorology. The emergence of the modern-day science of meteorology is continuously shaped by the changes in the philosophical understanding of nature. In the beginning the study of climate and seasons was primarily restricted within the larger ambit of Aristotelian philos-

[37] Frisinger, *The History of Meteorology*, p. 26.

ophy as explained earlier. However, further progress in this field was propelled by the necessity of prognostications for the contemporary societies.

One of the many scientists who made an important contribution in the field was Claudius Ptolemy, or Ptolemy the Astronomer (*c*. 85–165 AD). His *Almagest* is antiquity's most sophisticated astronomical treatise. In ancient times meteorology was generally considered a branch of astronomy; it is not surprising that Ptolemy displayed interest in weather phenomena and forecasting. In the *Tetrabiblos*, Ptolemy gave several astrological weather prognostications based on the appearance of the moon and other natural phenomena. We must observe the moon in its course three days before or three days after the new moon, full moon and the quarters. Ptolemy suggested that when the moon appears thin and clear and has nothing around it, it signifies clear weather. If it is thin and red, and the whole disk of the unlighted portion is visible and somewhat disturbed, it indicates winds in that direction in which it is particularly inclined. If it is observed to be dark, or pale, and thick, it signifies storm and rains.[38]

Similar sentiments can be deciphered in the writings of medieval Arab geographers who

> never had any concept of horizontal circulatory gyres in the ocean, and they always spoke of currents as water periodically emptying out of one place and piling up in another. Probably they owed this way of thinking to Aristotle, whose works were being translated into Arabic by the ninth century. . . .[39]

A similar understanding of the monsoon continued for a very long time with minor modifications. It will not be out of place to reiterate medieval categorizations of the world to situate the limited understanding of meteorological features. As discussed in previous chapter, a sixteenth-century court chronicler of the Mughal emperor Akbar, Abul Fazl, suggested that:

[38] Ibid., p. 27.

[39] The understanding of winds and currents was even less satisfactory, because the physics of the time was a muddle of Aristotelian natural philosophy, Islamic religious sentiments, astrology, and folklore. Even at its best the theorising did not lie in deduction from general physical principles, but in qualitative arguments from analogies, . . . Warren, 'Ancient and Medieval Records', p. 149.

The learned have divided the *oikoy* into seven parts, to each of which they have given the name of *Klima*. The Seven belts which these lines form are called climates. . . . There are two reasons for selection of seven as this number. The first is that ancient sages have verified by experience that each tract of superficial area was specially connected with one of the planets, as for instance, the first climate with Saturn. For this reason the inhabitants of that zone generally are dark skinned, curly haired, long-lived and indolent in action.[40]

Moreover, in accordance with the dominant belief system, certain amount of divine design was also associated with the cosmos in general and weather in particular. It was believed in Europe that lightning was one of the manifestations of divine moral force. In the Middle Ages and Renaissance, thousands of people, mostly women and children, were forced to confess to inciting hailstorms or deluge with sorcery. A woodcut in a 1519 edition of *Physics*, published in Augsburg, shows devils at play in the clouds, controlling the weather.[41]

Seen as part of a larger conception of the universe where celestial bodies were supposed to have influenced the climate and weather, it was but natural to witness growth in astronomical studies.

'We would then come across, for example, an outstanding astronomer, Mahendra Suri, in the reign of Firuz Tughluq. Suri wrote a treatise on the construction and use of armillary spheres, and the calculation of celestial and terrestrial longitudes and latitudes. Tughluq himself was a keen astronomer and is said to have build a number of observatories and also a clock, *Tas Ghariyal*, at his capital in Delhi. King Humayun (16th century) was again a keen observer of astronomical phenomena; he died as a result of a fall from an observational tower! During Akbar's reign extensive observations were carried out by Mulla Chand who compiled *Tashilat Mulla Chand Akbar Shahi*. During the Shah Jehan's reign we have two outstanding astronomers: Fariduddin Munnajam Dehlavi and Mahmud Jaunpuri. Shah Jehan, it appears wanted to build an observatory at Jaunpur but was, it is said unable to do so for lack of funds.[42]

40 Abul Fazl, *Ain-i-Akbari*, vol. III, pp. 50–3.

41 Frisinger, *The History of Meteorology*, p. 33.

42 Rahman, *Maharaja Sawai Jai Singh II*, pp. 15–16.

Improving meteorology in the time of the Renaissance, however, would have meant not only outsmarting Aristotle but also defying the Pope. In two areas, Aristotelian science failed miserably: terrestrial and planetary motion. The Enlightenment had brought many changes, but popular belief was more powerful than reason when it came to climate. Roger Bacon (1214–94) challenged the Aristotelian theories for their inhibition against experimentation. He stressed the importance of the verification of these theories in the real world. It was, however, Bacons insistence on the importance of experimentation and observation in science rather than on the authority of ancient writers which was Bacon's prime contribution to the development of meteorology.[43] It is important to recognize that most of the pioneering studies of the time were carried out without the benefit of meteorological instruments. Therefore, it is not surprising to note that the study of meteorological phenomenon by the scientists of antiquity was largely qualitative rather than quantitative.[44]

RUPTURES IN TRADITIONAL UNDERSTANDING

Ruptures in the conventional conception of the universe and thereby climate came from other regions as well. The greatest Moslem physicist and one of the most noted students of optics of all times was Ibn Al-Haitham (Alhazen) (*c.* 965–1039). The Latin translation of his main work, *Opticae Thesaurus*, exerted a great influence upon Western science, and represented considerable progress in the science of meteorology. In it Alhazen discussed atmospheric refraction and gave the first correct definition of twilight. He showed that twilight begins when the sun is 19° below horizon. Using this result, Alhazen attempted to measure the height of the atmosphere. In a very complicated geometric demonstration, drawing often upon Euclid, he arrived at a maximum height of 52,000 'passuum'—about 49 miles.[45] This was one

[43] Frisinger, *The History of Meteorology*, p. 33.

[44] Ibid.

[45] Ibid.

of the first explanations of climatic factors in terms of actual calculation based on the observed data and verifiable time. It marks the beginning of field observations.

The need and scope for field observation received a tremendous boost during the 'age of geographical explorations' in the fifteenth and sixteenth centuries. The discovery of sea routes to the oriental world and later on the discovery of the 'New World' unleashed entirely new dimensions on the conception of world geography.[46] At the same time the publication of Ptolemy's world map in 1482 once again argued that the earth is spherical. Moreover, Copernican revolution redefined the new science of motion pioneered by Galileo and Descartes—a theory that applied to all bodies, whether terrestrial or celestial. The time was ripe for a reassessment of the nature of science and its methods.[47]

Nevertheless, growing commercial interest across the globe and the resultant integration of different parts of the world led to the growth of meteorology along two divergent lines: the theoretical or pure science based on Aristotle's *Meteorologica*, and the applied pseudo-science of weather prognostications carried on by astrologers. Weather prediction by natural signs became very popular; astrometeorology enjoyed the protection of the church and of princes.[48] The commercial integration of the world indirectly facilitated the growth of a global network of observation systems which facilitated uniformity in the scale of measurement.[49]

The one man, however, who most typified both the end of the era of speculation and the birth of modern meteorology, was the 'first modern philosopher', Rene Descartes (1596–1650).

[46] Fernand Braudel, *Civilisation and Capitalism 15th–18th Century*, vol. III, *The Perspective of the World*, tr. Sian Reynolds, Collins, London, 1984, p. 139; Carlo M. Cipolla, *Before the Industrial Revolution: European Society and Economy, 1000–1700*, Methuen & Co. Ltd, 2nd edn., 1981, pp. 223–40.

[47] Stathis Psillos and Martin Curd, 'Introduction,' in Stathis Psillos and Martin Curd (eds.), *The Routledge Companion to Philosophy of Science*, Routledge, Abingdon and New York, 2008, p. XX.

[48] Frisinger, *The History of Meteorology*, pp. 35–6.

[49] Grove and Damodaran, 'Imperialism, Intellectual Networks.'

His *Le Monde* (The World or The Universe), composed in the early 1630s, was a treatise on physics and cosmology, which resolutely avoided the old Scholastic apparatus of "substantial forms" and "real qualities" and instead offered a comprehensive explanatory schema invoking only simple mechanical principals. A corner-stone of Descartes' approach was that the matter throughout the universe was essentially the same type; hence there was no difference in principle between "terrestrial" and "celestial" phenomena, and the earth was merely one part of a homogenous universe obeying uniform physical laws.[50]

In 1637, Descartes published his renowned book *Discours de la Methode*, in which he expounded his philosophy of scientific method. The basis of this method was fourfold:

1. Never accept anything as true unless one clearly knew it to be such.
2. To divide every difficult problem into small parts, and to solve the problem by attacking these parts.
3. To always proceed from the simple to the complex, seeking relationships everywhere.
4. To be as complete and thorough as possible in scientific investigations allowing no prejudice in judgement.

In an appendix to this work, *Les Meteores*, Descartes applied the above principles to his discussions on meteorology.[51]

The basic physical and mathematical tools needed in the study of the atmosphere were also extensively developed by 1800. The barometer and the thermometer were invented in Italy. Galileo invented the thermometer in 1607. Later, in his dying days, he suggested a tubular contraption that might prove the ever-present weight of air in the atmosphere. His assistant, Evangelista Torricelli, took him up on the idea and in 1643 invented the barometer. It did not take much longer for astute observers to link the fall of local air pressure to stormy weather. It was not by accident that Torricelli

[50] Ted Honderich (ed.), *The Oxford Companion to Philosophy*, Oxford University Press, Oxford and New York, 1995, p. 189.

[51] Frisinger, *The History of Meteorology*, p. 37.

wrote, in a letter to a friend in 1644, 'We live submerged at the bottom of a sea of elemental air'.[52]

> But thermometers were for many decades commonly exposed in an unheated north-facing room . . . and their calibration gave trouble because of the ageing of the glass. . . . Other problems in using instrument records lie in the miscellany of different units and scales used. (Sir Isaac Newton already in the 17th century had a thermometer scale that defined 0° as the freezing point of water, but took as its other fixed point the normal temperature of the human body, labelled as 12°.)[53]

It is true, of course, that little scientific use came of the new instruments before meteorologists organised networks of trained observers. But the future direction of meteorology was set.

A better comprehension of the geography of the earth, especially a growing belief in the heliocentric[54] nature of the earth, on the one hand challenged the conventional conception, and on the other hand, though vaguely, suggested interdependence of the climate of different regions. A synoptic vision of climate started growing but it was restricted by the inability to share observations quickly across different regions of the earth. The first step was, of course, the gathering of scientific data, which involved meteorological observations. There is little doubt that weather observations were made and recorded throughout antiquity. The Greeks, in fact, were one of the first to make regular meteorological observations from as early as the fifth century BC. The results were exhibited to the public on the so-called parapegna, a form of almanac fixed on columns.[55]

The importance of observed data for a better comprehension of the climate became evident very soon and attempts were made to encourage communities in other regions, nations to maintain the records observed with the help of instruments. It is interesting

52 Rosenfeld, Eye of the Storm, pp. 14–15.

53 H.H. Lamb, *Climate, History and the Modern World*, Methuen, London and New York, 1982, pp. 67–8.

54 I. Bernard Cohen, *Revolution in Science*, The Belknap Press of Harvard University Press, Cambridge, Massachusetts, London, 1985, pp. 105–45; Cipolla, *Before the Industrial Revolution*, pp. 240–7.

55 Frisinger, *The History of Meteorology*, p. 99.

to note that attempts to maintain a record of daily temperature can be seen at the global level. Protestant missionaries working in southern India were maintaining and communicating records of daily temperature.

We have made our observations on Patricks thermometer having everyday strictly observed ye whole month of December last (??) what observations we have made you find here set down in a particular paper whereby you may see the difference or degrees of heat and cold in the east Indies, and compare it with what you have in England. As soon as we come to the hottest day we do a sign to make another experiment of which we shall give you an account, with the first shipping. As for the uppermost glory filled with quicksilver, we do not know the use of it or how to manage it.[56]

Relevance of data to be maintained in different regions received further impetus with the discoveries of regional general circulation of the atmosphere with the distribution of the sun's heat over the earth's surface by Edmund Halley (1656–1742). He suggested that heated air rises, and that the winds were caused by air flowing into these areas where the solar heat had caused the original air to rise. Halley used this theory to explain the easterly trade winds around the equator. The sun exposed the equatorial zone to a great amount of heat everyday. The maximum heat follows the sun and, therefore, moves from east to west.

George Hadley (1685–1758) in his book *Philosophical Transactions*, for the first time made allowances for the earth's rotation. Hadley first depicted the atmosphere surrounding the earth as moving with it, so that there was no apparent motion of the atmosphere, and thus no wind. Then, he claimed, the action of the sun causes the air near the equator to be heated, rise and be replaced by cooler air from the north and south. This paved the way for the study of dynamic meteorology by nineteenth-century scientists.[57]

[56] I owe this information to Mahesh Gopalan, *Letter no. 28, Letters Relating to the Protestant Mission to the East Indies,* vols. I-VII (SPCK), vol. 1, 1710–16.

[57] Frisinger, *The History of Meteorology*, pp. 123–30.

Even in oriental societies, seventeenth and eighteenth centuries witnessed remarkable interests in attempts to unravel the systems of the climate in general and monsoon in particular. These attempts cannot be classified as pure scientific concerns as we understand today. Rather these were part of a larger process of documentation of geographic features. A conscious attempt on the part of the state to map the various geographical factors which directly or indirectly influenced agrarian production. Although following the conventional understanding whereby movements of astronomical bodies were considered responsible for climate phenomenon, seasons, etc., Jai Singh emphasized importance of accurate observation based on calculations and predictions.

> Jai Singh gives three main reasons for preoccupation with astro-nomical observatories. The first was the inaccuracy of the then existing calendars, namely the Gurgani, Ilkhani, and Ilahi calendars. The Gurgani calendar was the result of the efforts of Tusi, the Ilkhani was constructed by Ulugh Beg; and the last named by Mulla Chand during the reign of Akbar. Secondly since the "affairs of religion and administration of the empire" depended upon the correct calendar, there was a need to revise the calendar. Lastly, he points out the prevalent disagreements on the time of rising and setting of planets, the beginning and end of seasons, and the timing of eclipses of the sun and moon.[58]

Furthermore, it is significant to take note of the reasons for the construction of observatories by Jai Singh at different places.

> Why did he construct so many of them when one could have sufficed for making observations to achieve his purpose? There appear to be two reasons, as mentioned in his introduction. Firstly to make comparative observations so as to check and verify the measurements made at one place. Doubtless, a sound scientific approach. Secondly to promote the observational attitude amongst the people. . . .[59]

It is unfortunate to note that scholars have often ignored the long historical tradition of scientific developments, which often produced original works.

[58] Rahman, *Maharaja Sawai Jai Singh II*, p. 11.

[59] Ibid., p. 14.

These civilizations (Islamic and Indian) manifested a weakness in terms of absence of any scientific observation, systematization and communication of technological principles followed in the crafts. This is in contrast to what one begins to notice in the West European civilization from Agricola (early 16th century) onwards. . . .

Above explanation summarizes the conventional approach to science, scientific knowledge and developments in India.[60] Nevertheless, it is important that several attempts to develop observatories in India must be seen in the larger tradition of development of instruments.

The tradition of making instruments is also old in India. The first record we have is of Bhaskara (5th century) making a simple instrument. Then there is the development and construction of armillary spheres and possibly other instruments by Mahendra Suri at the time of Firuz Tughluq (14th century). . . . The family of Allah Dad Humayuni Lahori started making astrolabes from the time of Humayun (16th century) and continued to do so till the time of Jai Singh.[61]

Efforts of Sawai Jai Singh through construction of several instruments for astronomical calculations at different places should be seen in the larger context of central Asian and west Asian legacy.

He (Sawai Jai Singh) had acquired and studied the Gurgani Tables compiled by Nasiruddin Tusi at Maragha and the Ilkhani Tables compiled by Ulugh Beg at Samarkand and also other works and commentaries written by the scientists of the school of Ulugh Beg, like Kashi and others.[62]

Mere documentation of climatic data was not very helpful in large-scale predictions or predictions for a longer period. The dominant belief still visualized the climate in terms of Sir Isaac Newton's three laws of motion. His third law found greater applicability in the studies of climate and these were compatible with the local regions where variance in temperature and atmospheric pressure could be explained locally. Since nature was seen as

[60] Irfan Habib, '*Medieval Technology Exchanges Between India and the Islamic World*, Viveka Publications, Aligarh, 1985, p. 25.

[61] Rahman, *Maharaja Sawai Jai Singh II*, p. 16.

[62] Ibid., p. 17.

a machine, climate was explained in terms of mechanical processes of a machine which could go on forever.[63] What propelled this 'machine' could not be explained then, since nature was still conceived as a self-regulatory machine perhaps powered by unseen forces. The role of the sun as an agency of 'heat' was recognized but the actual mechanism through which it disseminated heat in the atmosphere was not understood.

This question remained a puzzle till the first law of thermodynamics was formulated in the middle of the nineteenth century. This law forms the backbone of meteorology. The motions of the atmosphere are caused due to thermal forcing and they are a thermodynamic phenomenon. Integration of the thermodynamic properties of elements to explain the motions in the atmosphere really emancipated meteorology from the traditional dogma. Nevertheless, at the same time it added further complexities to the science of meteorology. A fresh explanation was necessitated to explain the dynamics of the climate which were in consonance with the general principals of thermodynamics. A comprehensive explanation of climatic phenomenon was at times difficult to arrive at even in regional terms. Moreover, greater availability of data from different parts of the world combined with developments in the science of meteorology helped in formulations of theories about the climate as systems.

A major step forward was taken by Vilhelm Bjerknes. He suggested that with enough information about the current state of the atmosphere, scientists can use math formulae to predict future weather patterns. Unfortunately, due to lack of processing facilities (computers), the calculations he envisioned were not feasible at that time. Nevertheless, Bjerknes considered weather forecasting as the principal objective of meteorological research.[64]

Another major change in meteorology came with Bjerknes's insistence on aerological observations. He said that:

[63] Daniel B. Botkin, *Discordant Harmonies: A New Ecology for the Twenty-First Century*, Oxford University Press, New York, 1990.

[64] Robert Marc Friedman, *Appropriating the Weather: Vilhelm Bkerknes and the Construction of a Modern Meteorology*, Cornell University Press, Ithaca and London, 1989, p. 54.

by restricting chiefly to observations from near the surface, meteorologists had not progressed much beyond "the old weather seers", for both had to rely on external changes. The laws by which atmospheric conditions develop and change had consequently largely evaded meteorologists.[65]

Nineteenth-century forecasters witnessed frequent surprises and errors in estimating surface atmospheric pressure patterns. They believed that this was because of insufficient information about the upper atmosphere. The British meteorologists made a series of ascents by balloon during the 1860s. Soon, the era of aviation industry was emerging and was also seeking insights from meteorology to combat the uncertainty of the climate in their regular operations. This mutuality proved tremendously beneficial for the development of meteorology. Now it could venture into climate forecasting. Immense growth in the observation stations and data available ultimately culminated in small-scale predictions for local areas.

Though, in principal, these data offered possibilities of predictions at regional levels, but, given the magnitude of data, it was difficult to generate regional predictions. At this juncture:

> the device with the most profound impact and capabilities of meteorology was the computer, and not just because computers now make storm forecasts or process Doppler data from radars and generate images from satellites. Almost by itself, the computer has expanded meteorology from its humble beginnings as an observation-only science. With a few exceptions (such as precipitation experiments), weather has defied the controlled conditions of a laboratory that confer on physics, chemistry, and biology a special status.[66]

Along with the special characteristic that it defied the controlled conditions of a laboratory another important characteristic of global tele-connection of the climate could not be figured out till very recently. During the early 1920s, Gilbert Walker empirically identified a periodic variation in atmospheric pressure over the Indo-Pacific which he christened 'Southern Oscillation'. However, it was during the 1960s that Bjerknes posited a physical mechanism

[65] Ibid.

[66] Rosenfeld, *Eye of the Storm*, p. 13.

to explain the atmospheric features of this phenomenon over the equatorial Pacific and laid the path for global climate systems. Though, by the 1970s general circulation models (GCM) had become the central tool of climate science, they became really global when during the mid-1980s oceanic general circulation models (OCGM) were coupled with atmospheric general circulation models (GCM). With the development of general circulation models (GCM) it was expected that now climate could be predicted on a global level as even the global tele-connections were less or more now in place.

LITERARY DESCRIPTIONS

It should be emphasized here that for the construction of social history of any region and period, the use of literature could prove very useful as literature stands in recognizable relation to life. Lucien Febvre has explained this basic relationship in the following words:

> Intellectual activity (literature) presupposes social life. Its essential instruments (with language in the forefront) imply the existence of a human society in which they had necessarily to be worked out, the aim of such instrument being to make it possible to set up relationships between all those who participate in one and the same environment.[67]

This, however, does not imply that literature is a mirror reflection of life. One has to exercise caution in using literature as a source for the study of society. The literature in fact moulds and constructs reality, making people perceive patterns hitherto unknown to them. Our endeavour in this work will be to construct social life with the cautious use of literature.

Seasons are important manifestations of environmental features. These have been represented in the contemporary literature. The literary sources broadly highlight the social perceptions of the environment. The similies are chosen from the immediate physical context and from the social environment. We have cited a few of

[67] Peter Burke (ed.), *A New Kind of History: From the Writings of Lucien Febvre*, tr. K. Folca, Kegan Paul & Company, London, 1979, pp. 12–27.

such excerpts. For a better and comprehensive analysis of texts, similar categories have been clubbed together.

A section from Bargawat Dev Narayan's folk tales very clearly highlights the significance common man attaches to the rains. The conversation between various characters points out the common concerns.

When asked by a character, Hira, as to which is the most important flower, another character, Niya, replies,

> Cow! Because it is the basis of economy of the region. Its calf is used in agricultural works. A great flower is the horse, which forms the mainstay of the kingdoms. A great flower is cotton, whose derivative is clothes which cover the world. A great flower is the sun, for the dawn augurs the start of the day's work. However, the greatest flower is rain, for if there were no rains, all other flowers would not blossom. Rains are the mainstay of life in the region.[68]

Similarly, Manikya Sundar Suri, a fifteenth-century poet praises the rainy season in his work *Prathivi Chand Carita*:

> इसिइ अवसरि आविउ आषाढ़, इतर गुणि संबाढ ।
> काटइ यइ लोह, धम तणउ निरोह ।
> छासि खाटी, पाणी बीयाइ माटी ।
> बिस्तरिउ वर्षाकाल, जे पंथी तणउ काल ।
> नाठउ दुकाल, जीणिइ वर्षाकालि ।
> मधुर ध्वनि मेह गाजइ, दुर्भिक्ष तणा भय भाजइ ।
> जाणे सुभिक्ष भूपति आबतां जय ढक्का बाजइ ।
> चिहुंदिसि बीज झलहलर, पंथी घर भणी पुलइ ।
> विपरीत आकाश, चन्द्र सूर्य परियास ।
> राति अंधरी, लबइं तिमिरी ।
> उत्तरमउ उफनयण, छायउ गयण ।
> दिशि घोर, नाचइं मोर ।
> सधर, बरसइ ध्राध्र ।
> पाणी तण प्रवाह खलहलईख बाड़ि उफ्परि बेला बलइ ।
> चीबलि चालतां षकट सखलइं लोक तण मन धर्म्म ।

[68] Lakshmi Kumari Churnawat, 'Bargawat Gatha: Itihas ke Jharokhen Se', in Jawahar Lal Handoo, (ed.), *Folklore of Rajasthan*, Central Institute of Indian Languages, Mysore, 1983, pp. 83–4.

उफपरि बलइ। नदी महापूरि आबई, पृथ्वी पीठ प्लाबइं ।
नवां किसलय बल्ली लहलहइ ।
कुटुंबी लोक माचइ, महात्मा यइठां पुस्तक बांचइं ।
पर्वत तउफ नीसरण बिछुटइ, भरियां सरोवर पूफटइं ।[69]

It says that now Ashadh, the rainy season has come. It has got immense qualities. With the coming of rains, one gets relief from extreme heat. The thirsty soil gets water. It is like a traveller, which had forgotten the route to this region. Its advent precludes drought. Long live the rainy season. With the sweet sound of the clouds' movement, the fear of famine is demolished. It ensures good harvest so the king is cheerful and plays drums. All around there is greenery, which will ensure good harvest so that the peasants' granary shall be packed. When the sky is unfavourable, the sun and moon are shining bright, nights are dark and long, we witness the emergence of clouds in the north and covering the whole sky; peacocks are dancing, the rains continue to pour. Water will flow with great speed and flowers will blossom. Rivers shall overflow, land shall submerge. Creepers will flourish. People shall stage dramas and the learned shall indulge in writing. Insects shall migrate to mountains, upland, and ponds shall overflow.

A similar perception is visible in one of the sections written by Yati Vradhichand ji in the sixteenth century as *Mulkalanupras*, which is available in *Pakaik Vishanti*, a text written by the seventeenth-century poet Sur Chandra:

आषाढ़ साढू मेघ आया, कुणइ कनइ मनि उछरंग न भाव्या
कालहणी बली, सर्वजन नइ मन रली ।
उत्तर वायु बाज्या, आकासि मेधि गाज्या ।
कुड़ा बहक्या, केवड़ा महक्या ।
कुंद उल्लस्या, करसणी हरिस्या ।
बदंब महमह्या, मयूर गह गह्या ।
बप्पीहा बासइ, बिरहणी उफसासइ ।
पर्वत शिखरि, स्नेह नइ भरि ।
सोगलुं वायइ, मल्हार गायइ,
भील नाचइ, महिषी माचइ ।

[69] Agarchand Nahata (ed.), 'Rajasthani Gadh Main Varsha Varnan', *Maru Bharti*, vols. I–II, 1952–5, pp. 63–4.

बूठा मेह, उल्लस्या स्नेह,
नदी महापूरि बहिबा लागी, पगन लहई पागी ।
जलस्यू भर्‍या निवाण, पृथिवी प्रवर्त्ती मदन भी आण ।
हरी प्रगट हूया, दीसइ बराह राजूथ जूजुया ।
सालूर ना सांभलियइ स्वर, जाइ दीसइ विकस्वर ।
भलां केलबियइ बालर, वावीयइ झालर ।
अति रूप, नीपजइ झूप ।[70]

With the coming of Ashadh the clouds have come, there is nobody who is not celebrating it. Everyone appreciates the heavy rains. Northern winds have brought the clouds, which have covered the sky. The fragrance of *kewra* has enveloped the land. The flowers of the pond have inflorescence, peasants' wives are overjoyed. The *kadamba* tree has blossomed. Peacocks are dancing. The *papiha* has started sorrowful cries; the *virahani* bemoans. Mountain peaks have once again become green. *Seeglun* has started blowing. The song of rains, *Malhar*—a raga—is sung everywhere. *Bhils*, tribals, are dancing. Saints are studying. The clouds have brought joy to everybody. Rivers have started overflowing. Footprints have covered the region. Water has been accumulated in the ponds, and land has become the abode of Madan—the god of love. God has been kind; clouds have spread over the area. The cries of frogs can be heard everywhere. Deer have started roaming around. Step wells are full of water. Trees of lemon are full of good fruits. The place is full of trees of Kair.

A section from the seventeenth-century *Kalikacharya Katha* represents rains in the following terms:

आवउ वर्षाकाल, चिहुंदिसि घटा उमटी ततकाल ।
गड़ढ़ाट मेह गाजै, जाणै नालि गोला बाजै ।
कलै आभै बीजुली झबकइ,विरहणी ना हीया दबकइं ।
बब्बोहा बोलई, वाणिया धन बेचिबा बारबार खोलइ ।
बोलद मोर, दादुरई सोर। अंधरइ घोर, पइसइ चोर ।
कंदर्प्प करइ जोर, मानिनी स्त्री भर्त्तार नइ करइ निहोर ।
चंद सूरिज बादले छाया, पंये बाउफ आप आपणां धरां नइ
धया राजहंस मानसरोवर भणी चाल्या,
लोके वस्तु बाा धरां मांहे धल्या ।

[70] Ibid.

बग पंकति सोहइ, इंद ध्नुष चित्त मोहइ ।
आभौ थयौ रातौ, मेह थयौ मातौ ।
मोटी छांट आबइ, लाकां नइ भावइ ।
झड़ी लागी, करसणी री भाग्य दसा जागी ।
मूसलधर मेह बरसई, पृथ्वी उर्ण-पूर्ण करिया तरसइ ।
बहइ प्रनाल, खलखलइ पाल ।
चूयई उरा, भीजइ वस्तुबो ना बोरा ।
अबकइ पटसाल चिचुयइ बाल ।
नदी आबी पूर, कडणि रा रूंख भोजि करइं चकचूर
बहइ बाहली, लो थया काहला ।
जूना ढूंढ़ा पड़इ, लोक ऊंचा चड़इ
हाली ए खेत्रा खड्या, बाड़ि स्यु सेढा जडया ।
मारग भागा, जे जिहां ते तिहां रहिबा लागा ।
प्रगटया राता मामोला, धन थया सुंहगा मोला ।
नीली हरी डहडही, धणाथया दूध नइ दही ।
नीपना घणा धन, संभर या धर्म नइ धयांन ।
गयो रेर, लोग करइं बकोर ।
गयो दुकाल, थयो इंइं सुगाल ।।[71]

With the coming of the rainy season clouds have engulfed the whole sky. The thundering of the clouds is like the sound of gunshots. At times lightening is taking place, thereby threatening the *virahani*—lonely beloved. The *papiha* has started crying and merchants are liberally selling their goods, as they are sure about a good harvest season. Peacocks are dancing. The sky has become dark, encouraging thieves. Desires are running high, and lovers are engrossed in the beauty of their beloved. Clouds cover the sun and moon. Swans are playing in the ponds. The groups of heron are looking pretty. The rainbow is very beautiful. The day has become night due to clouds. The raindrops are big in size and the whole scenario has changed for the good. Rains have brightened the future of peasants' wives. Rains are heavy, satisfying the thirst of the land. Drains shall overflow; embankments shall be full. The land will receive good water and things kept inside bags shall be drenched. The water shall leak in the stable, and drench the horsetails. Rivers are flowing full, carrying with them trees, etc. *Bahlas* are flowing, giving immense pleasure to the people. Old houses have fallen down, people have moved to higher

[71] Ibid., pp. 64–5.

places. Peasants have ploughed their land. Owing to the floods the people migrate and live wretchedly. People are enjoying the abundance of milk products. The cultivation of grain has been very good. Now one can devote time to religion. The period of hardships is over, there shall not be famine and the future shall be peaceful and prosperous.

Sam Shringar, a text written in AD 1762 depicts rains as:

ऊमटी घटा, बादल हुवै एकठा पड़इ छ्टा, भाजइ भटा
भीजइ लटा, पुहवी पुणय प्रगटा
कषि राजान ठांस बैठा मेह गाजइ,
जाणै नाल गोला बाजै
दुकाल साजै, सुवाय बाजइ
इद्र राजै दाय पराजै
बीजली सबकै मेह टबकै घणा
जीवरी उपपात कोइ पंथ चालउ मत ।
बोलइ मोर डेडक चोर ।
अंधरो घोर, पैसइ चोर
भीजइ ढोर, स्त्री करइ निहांर
चंद सूर बादल छायौ, पंथी घर बहै आयौ
मेघ बरसै सबायौ, रूठौ नाह मनायौ ।
खलकै लाख नदी बहै असराल, बहै प्रणाल
चूंचे बाल, चूइंउ रासाल, साप पायाल ।
झड़ी लागी, करसणा री दिसा जागी
बरसणो लागौ पूर, भांजैरूख चकभूर
हाटा बिचै याहला, लोक थया आकला ।
जूना घर पड़ै, लोक ऊँचा चडै
आभा हूआ राता, मेह थया माता ।
हालीइ हल खड्या
नीली हरी महमही, घणां दूध नै दही ।
मारग भागा, जिहा तिहां बैसवा लागा ।
पाणी छाडै पाल, गयौ दुकाल ।
हूओ सुगाल, पुहवौ वर्षाकाल ।[72]

Rains have come, so clouds have appeared. It is very beautiful, The Bhat (Bards) are singing songs. Creepers are drenched. It is definitely due

[72] Ibid., p. 65.

to good deeds. The thunders of clouds are like the sound of gunshots. Last year had been a famine year. The god of rains has come to our place. Lightning accompanies the rains; one should not venture to go out. Peacocks are singing. Frogs have come out. It has become dark and thieves are on a prowl. The cattleherds are overjoyed and the ladies are jubilant. Clouds have covered the sun and the moon; the peasants thus come back early. The rains should be so heavy that rivers should flow full with great water, *bahlas* should flow. The rains have been god's gift to the peasants. Rivers are flowing full, carrying with them trees, etc. *Bahlas* are flowing, giving immense pleasure to the people. Old houses have fallen down, people have moved to higher places. Peasants have ploughed their land. People migrated and started living here and there. People are enjoying, the immense production of milk products. The cultivation of grain shall be very good. Now one can devote time to religion. The period of hardships is over, there shall not be famine and the future shall be peaceful and prosperous.

Similarly:

कौतूहलम् में
उमटी घटा, बादल होइ एकठा ।
पड़इ छ्टा भाजइ भटा भीजइ लटा
मेह गाजइ जाणो नाल गोलावाजइ ।
दुकाल लाजइ सुवाब वाजइ
इन्द्र राजइ, ताप पराजए
बीज झबकै, मेह टबकै
हीया दबकै, पाणी भमकै
नदी उबकै, बनचर लबकै
आयो अबकै,
बोलइ मोर, डेड करै शोर
अंधर घोर, पइसइ चोर
भीजै ढोर, खलकै खाल, बहै परनाल
चूयै साल, साप गया पयाल
झड़ लागी, लोक दसा जागी
घर पड़ै, लोक ऊंचा चडै
आभा राता, मेह माता[73]

[73] Ibid., pp. 65–6.

Rains have come, so clouds have appeared. It is very beautiful, Bhat (Bards) are singing songs. The thunders of clouds are like sound of gunshots. Famine shall not visit this year. It will be a good year. The god of rains has come to our place. Heat will recede. Seeds will be sown, as rains will continue. Everybody will celebrate. Rivers are flowing full. Animals in the forest are enjoying it. Peacocks are singing. Frogs have come out. It has become dark and thieves are venturing out. Livestock is also enjoying rains. *Parnala* should flow. People are enjoying. Old houses have fallen down, people have moved to higher places.

In the semi-arid and arid regions of Rajasthan monsoon failures would lead to drought-like situations and continuous visitations of drought would lead to famine. The rainy season was thus welcomed as it ensured better agricultural prospects and potable water supply for the forthcoming year. Despite hardships caused by rains, the rainy season was heartily welcomed. Even though rains may force them to vacate their home and force them to shift to higher places for safety against flood, yet rains were welcome. It ensured good times and eliminated possibilities of draught or famine.

Like any other monsoon-dependent economy Rajasthan had been prone to frequent droughts and resultant famines. The failure of rains was the primary cause of droughts and delay in the arrival of rains caused disruption in the production process, at times causing famines. Various methods of the prediction of rain were prevalent, so as to enable them to take preventive methods. Numerous rituals were prescribed to ensure good rains, which were practised at the level of individuals, villages and also at the state level. The *arzdashts* provide evidence for the various rituals and superstitions, which were performed under the instructions of the state to ensure timely rains. Rituals like the worshipping of a *vat* tree or organizing a *Brahma Bhoj*,[74] *Sava Lakh ka Jap*[75] prayers involving Brahmins and the *qazis*, were held. At another level it was a common practice to construct *talabs*, tanks, *baori*, etc., in

[74] *Arzdasht*, Sawan Sudi 6, 1774, vs, HS, JR, RSAB.
[75] *Arzdasht*, Bhadva Vadi 5, 1762, vs, HS, JR, RSAB.

one's name for public use. Such a deed had religious connotations also.

It is therefore interesting to note that the monsoon in traditional societies, as it is even today, was a cause of concern. If the timely onset of the monsoon generated possibilities of plenty in agrarian production then delay or failure prompted initiation of remedial measures. Human–monsoon relationships in the past have been very thick and were manifested at multiple levels as argued here. Social perceptions, scientific understandings of the monsoon broadly outlined the concerns of humans in the arid and semi-arid landscape of Rajasthan and the delineations in this chapter uncover a part of this relationship. In the later chapters a discussion of the human adaptation to the climatic variability and vagaries of the monsoon will try to encapsulate the multiplicity of human–monsoon interactions.

THE RAIN-CATCHERS

CHAPTER 3

Fluid Stability
Variable Waterscape and Settlement Patterns

It is necessary to appreciate the fact that the dynamics inherent in human–waterscape interactions, especially in monsoon ecologies, more often than not generate continuous negotiations and renegotiations. In pre-modern times, a unidirectional growth of societies was possible in cases where long-term flux in waterscape was minimal. Societies were usually able to cope with the short-term flux. Long-term flux often eroded the sustenance base of these societies.[1] Changed waterscapes necessitated new negotiations which at times culminated in the inability to sustain large urban set-ups and often resulted in migration to the agriculturally marginal areas or the emergence of new patterns of settlements. Given the nature of historical sources, coupled with limited excavations, these negotiations have often been, unfortunately, examined in

[1] Almost all civilizations take sufficient precautions to tide over short-term climatic variability or the flux but were usually unable to sustain themselves during long-term climatic variability. Evidence of elaborate water-works testifies to the fact that societies were aware of climatic variability and tried their best to manage it. However, long-term climatic variability was difficult to manage unless changes in the resource use pattern were initiated and executed. We also have evidence of the process of adaptability to the long-term climatic variability from almost all the human settlements and civilizations. Please refer to Fagan, *The Long Summer; The Great Warming*; Jared Diamond, *Guns, Gems and Steel: The Fate of Human Societeis*, W.W. Norton, New York, 2005.

terms of the fate of dynasties. Examinations of social negotiations of the human–waterscape interface have not been accorded priority. Recent writings on the environmental history of India have centred on the examination of 'negative feedback mechanism' and barring a few exceptions have mostly been concerned with the colonial or post-colonial period as has been pointed out in the introduction. At the same time, these writings tend to assume a particular image of the pre-colonial state, as one that was distant from the rural agro-ecological system. Nonetheless, it will not be inappropriate to state that adherence to stationarity[2] is clearly visible in the examination of the human–nature interface. The first section of this chapter focuses on the north-western region to examine in detail the nature of climatic variability. It is followed by a critical evaluation of the processes of adaptations to bring out the inappropriateness of notions of stationarity, in the larger context of monsoon ecologies. The next section elucidates the historical process of the settlement of different parts of Rajasthan while negotiating monsoon-dependent irrigation for agriculture, and greater reliance on agro-pastoral sustenance. The last section examines the implications of integration with the larger agrarian landscapes of the Indo-Gangetic plains.

Usually, socio-economic profiles of societies have been analysed in terms of production relations where the waterscape has been placed in the background.[3] Most of the studies have at the most relied upon written sources to examine the complex interactions between social formation and landscape; the waterscape being a minor component of the landscape. Unfortunately, archaeological sources have only been allowed a secondary role, often entangled with the debate focusing on the fate of larger social formations. Though 'marginal' in terms of political formations, at least till the early medieval times, Rajasthan and adjoining areas of Sind and Sistan offer possibilities of a closer examination of the

[2] P.C.D. Milly, Julio Betancourt, Malin Falkenmark, Robert M. Hirsch, Zbigniew W. Kundzewicz, Dennis Lettenmaier and Ronald J. Stouffer, 'Stationarity Is Dead: Whither Water Management', *Science*, 319, 2008, pp. 573–4.

[3] Mukhia, 'Was There Feudalism', pp. 44–53; see also D.N. Jha (ed.), *Feudal Social Formations in Early India*, Chanakya Publications, Delhi, 1987.

human–waterscape interactions.[4] *Prima-facie* these studies accept the changes in the waterscape but did not probe the process of adaptations to these changes. Thus, it will be interesting to examine the complex process of adjustments and accommodations carried out by humans to cope with the changed waterscape. There is a possibility that a fresh look may unsettle already set categories and may not necessarily at the same time offer well-defined fully crafted alternative categories. But this does not mean that the fresh look should be ignored simply because it is not able to totally dislodge already settled categories. 'The point is: if there is need to look at change in early Indian history, it is necessary that historians too occasionally introduce some change in the way they see the past.'[5] Therefore, it is argued here that an examination of the process of social negotiations as a process may offer us a different understanding; an understanding in terms of adaptability and flexibility.

Directly or indirectly, the monsoon has been the primary author of the irrigation and agriculture scenario for the whole of India. Apart from the tectonic movements, monsoons have been the most dominant factor to influence the changes in landscape witnessed over the Indian peninsula. It will not be an exaggeration to argue that the human–nature interactions in north-western Rajasthan and adjoining areas are dictated by the monsoons. Winter rains have been more or less insignificant. It is not only the regular annual cycles, but the timing of the onset, duration, intensity and the gap between two successive showers, that influence agricultural production over vast areas of the Indian peninsula. Variation in all or in any one of these factors resulted in continuous adaptations by

[4] B.D. Chattopadhyaya, 'Irrigation in Early Medieval Rajasthan', *Journal of the Economic and Social History of the Orient*, vol. XVI, pts. 2-3, December 1973, pp. 298–316; 'Origins of the Rajputs: The Political, Economic and Social Processes in Early Medieval Rajasthan', *Indian Historical Review*, 3, 1976, pp. 59–82; 'State and Economy in North India: Fourth Century to Twelfth Century', in Romila Thapar (ed.), *Recent Perspectives of Early Indian History*, Sangam Books, Delhi, 1995, pp. 309–46.

[5] Chattopadhyaya, 'State and Economy', p. 337.

humans.[6] As discussed in the Introduction, despite advancements in the field of paleoclimatic studies the construction of past climate variability remains inconclusive. Most of the formulations are highly debated and there is no consensus among scientists.

CLIMATE VARIABILITY

It is not surprising to note that any investigation of the past or present climate/s of the Indian subcontinent will most probably begin with an appraisal of climate in general and the monsoon in particular. As has been argued in Chapter 2, the major impetus in understanding the variability of the monsoon came very recently.[7] Now it is generally accepted that the monsoon is a regional phenomenon and intricately integrated with the global tele-connections. Though recent studies have broadly established possible links among numerous regional climatic patterns to account for the nature of the monsoon in India, the most prominent of these being the close connection between El Nino and the Indian monsoon; nevertheless, it is important to recognize that:

> there is no single, closely definable global footprint of El Nino's tele-connections, owing to the existence of oscillatory climatic systems in other oceans, which to a degree act independently of El Nino. Apparently, statistical correlations between El Nino and droughts in the Ethiopian highlands, India, and Java, for example are strong but patterns differ substantially between individual cases, as illustrated by the 1997–1998 events, which was accompanied by severe droughts in Southeast Asia, whilst the South Asian monsoon was barely affected. Again, hard winters and times of poor crops in Europe sometimes coincide with El Nino, depending on the phase of the north Atlantic oscillation. Furthermore, the forerunner signs of a global El Nino episode may lie outside the Pacific basin; indeed, failure of the South Asian monsoon sometimes preceded the eastward translation of warm surface water in the pacific or the advent in north Peru of the El Nino, rather than coincide with it.[8]

[6] Kumar, 'Situating the Environment'.

[7] Das, *Indian Monsoon*.

[8] Grove and Chappell, 'El Nino Chronology', p. 6.

It is also important to realize that the climate is an extremely complex system and we must be very cautious while drawing conclusions. The limitations of global tele-connections to establish direct correlations between ENSO (El Nino Southern Oscillation) and the drought (failure of the monsoon) were once again raised by R.K. Yadav who suggests that there may be other factors to explain the onset of the monsoon.

> Despite the importance of the ENSO (El Nino Southern Oscillation) phenomenon in climate variability on the seasonal-to-inter-annual timescales, it does not explain all the variance of the Indian monsoon. For example, only about 50% of droughts in India are associated with ENSO. Therefore, some part of inter-annual variability may be influenced by some other part of the climate system, for example snow-cover variation over Eurasia.[9]

Nevertheless, it is safe to argue that this region has a history of climatic variability. There are evidences which suggest that even during the later period variability in climate can be located for this region.[10]

It is now a generally accepted fact that the climate during the Holocene era had witnessed enough variability and it cannot be described as stationary, not even in its periodic cycles.[11] The climate of Rajasthan has seen great variability ever since the period of human occupation. It has been argued that the Rajasthan desert was once the centre of a flourishing civilization.[12] Palaeontological

[9] R.K. Yadav, 'Changes in the Large-scale Features Associated with the Indian Summer Monsoon in the Recent Decades', *International Journal of Climatology*, no. 29, 2009, p. 117.

[10] Dhavalikar, relying on the assumption of global tele-connections as manifested in ENSO events visible on the Nilometer argues that: 'In the four centuries from 600-1000 AD, the Nile levels were extremely low for some 128 years, more particularly during 687-696, 770-782, 788-802, 945-951 and 963-67 AD. It seems that 770-802 was the long phase of low Nile levels, and we may not therefore be far off the mark if we infer that these were the years of severe droughts in India.' Dhavalikar, 'Green Imperialism', p. 26.

[11] Madella and Fuller, 'Palaeoecology and the Harappan Civilisation'.

[12] Examples of areas which are now deserts, but which at some time in the past enjoyed a favourable climate are not difficult to find. The climate of the Sahara was reputed to be wet and moist between 60,000 and 600 BC. Geological

studies by G. Singh and Reid Bryson suggest that the climatic history of the larger area of the Thar Desert and the adjoining region may be divided into four periods beginning with 8000 BC when the region was moist, wet and had a cool climate, which changed between 8000 and 3000 BC and a dry climate, but less arid than at present, is postulated. It was during this era that there are evidences for the beginning of agriculture in the region. This was followed by a period of higher rainfall between 3000 and 1700 BC when the urban civilization of Harappa flourished. C. Ramaswamy has suggested that:

> during the south-west monsoon seasons around 2000–500 BC deep troughs in the upper westerlies must have been extended into west Pakistan far more frequently than they do now, inducing monsoon activity and also causing any monsoon depressions from the Bay of Bengal moving towards longitude 78° E, to curve to the north or north-east.[13]

On the basis of the drying of fresh water lakes, dry conditions have been suggested for the period between 1700 and 1500 BC. Following the last period, the indications are that very arid conditions prevailed over Rajasthan.[14] Based on the historical and archaeological evidences Ghosh also corroborates it and notes that by the time of the composition of the *Mahabharata*, that is, roughly the early centuries AD, the desert already existed as far north as Bikaner.[15] Variability associated with the monsoon continued even for later periods and it has now been generally accepted that the Indian subcontinent witnessed a decline in precipitation and long eras of relatively less rains between the eighth and tenth centuries AD. Devra has pointed out that similar climatic variability can

---

evidence suggests that sometimes after the glaciers of the last ice age began to recede, the climate of the Sahara became increasingly hostile. Das, *Indian Monsoon*, p. 151. Interesting details on climatic variability and human settlement have been documented in terms of the genetic profile of humanity. See http://www.bradshawfoundation.com/journey (accessed 9 March 2010).

[13] C. Ramaswamy, 'Monsoon Over the Indus Valley during the Harappan Period', *Nature*, no. 217, p. 629.

[14] Das, *Indian Monsoon*, p. 150.

[15] Ghosh, 'The Rajputana Desert', p. 98.

be argued even for the adjoining region of Sistan. 'After seventh century, it drastically declined. Scientists argue that poor, uncoordinated management and excessive extraction of water for agriculture purposes, combined with long years of drought, has led to drastic declines in the water flows of Helmand River.'[16]

Furthermore, this period has been deciphered to have witnessed visible climatic variability. Dhavalikar, relying on the assumption of global tele-connections as manifested in ENSO events visible on the Nilometer argues that:

> In the four centuries from 600–1000 AD, the Nile levels were extremely low for some 128 years, more particularly during 687–696, 770–782, 788–802, 945–951 and 963–67 AD. It seems that 770–802 was the long phase of low Nile levels, and we may not therefore be far off the mark if we infer that these were years of severe droughts in India.[17]

It is equally important to point out that even the other regions of the Indian peninsula witnessed noticeable fluctuations in the climate during the early medieval period.[18] Dandin, the author of *Dashakumara-Charita* gives us an eye-witness' account of harrowing conditions in the city of Trigarta due to severe famine.[19] The *Vayu Purana* refers to a curse which made the city of Varanasi desolate.[20] Varahmihira, writing his *Brihat Samhita* in the latter half of the fifth century and the earlier part of the sixth century, predicted the downfall of several cities in north and south India including Taxila, Mathura, Ujjain, Mithila, Tripuri, Varanasi, Avanti, etc. He predicted evils days for the people of Gandhar. In

[16] Ghosh, 'The Rajputana Desert'. G.S.L. Devra, 'Environmental Crisis and Social Dismemberment in Northwest India During the pre-colonial period', paper presented at the *Public Lecture Series*, Nehru Memorial Museum & Library, New Delhi on 4 November 2011, pp. 1-16.

[17] Dhavalikar, 'Green Imperialism', p. 26.

[18] Deep Narayan Pandey, Anil K. Gupta and David M. Anderson, 'Rainwater Harvesting as an Adaptation to Climate Change', *Current Science*, vol. LXXXV, no. 1, 10 July 2003, pp. 46–59.

[19] M.K. Dhavalikar, 'Culture-Environment Interface: A Historical Perspective,' Presidential Address, Archaeology, Numismatic and Epigraphy Section, *Proceedings of Indian History Congress*, 52nd Session, Delhi, 1992, p. 16.

[20] Ibid.

sum, the north and central India during the post-Gupta period were in decay whereas Western India and the Deccan were facing severe droughts.[21]This is largely corroborated by archaeological evidence. A large part of north India does not offer evidence of continuity of settlements during the post-Gupta period.[22] 'Even in central and western India and the Deccan there is a break in the habitation after the Satavahana period and the sites are inhabited only in Medieval period and the same story repeats in Karnataka, Andhra Pradesh and Tamil Nadu.'[23] Climate variability and generally accepted enhanced aridity in the region are considered as factors initially reflected in the general desiccation and gradually extension of desert conditions towards the north and north-east.[24] The *Vamana Purana* states that several holy places were suffering because of the advance of sand dunes.[25]

Generally, the period has also been portrayed as a period of decline and decay. However, this view has been challenged by historians. Nevertheless, the decline of urban centres during this period cannot be ignored though it should not be equated with the general agrarian decline.

> The decline of ancient towns in early medieval times cannot be considered an indicator of decline of overall economic growth. What emerges now is a new type of economy marked by urban contraction and agrarian expansion. Urban decline and stagnation resulted in agrarian expansion, which was promoted by land grants made by chiefs and princes. The emergence of numerous states in the early middle ages is a significant pointer to agrarian expansion. Nearly fifty states existed in the fifth to the seventh centuries, and many of them appeared in such areas as had never experienced any regular state machinery.[26]

[21] Ibid.

[22] Sharma, *Urban Decay in India*, p. 178.

[23] Dhavalikar, 'Culture-Environment Interface', p. 18.

[24] Ghosh, 'The Rajputana Desert'; this narration has been challenged and a continuum of habitation and state formation in the Ghaggar–Hakra region has been documented by Devra, 'Problems in the Delimitation'.

[25] Dhavalikar, 'Culture-Environment Interface', p. 16.

[26] R.S. Sharma, *Economic History of Early India*, Viva Books, Delhi, 2001, p. 150.

Moreover, seventh century CE onwards, one can see the emergence of settlements in the agriculturally marginal areas increasingly relying on rainwater harnessing available primarily during the monsoon.

Moreover, it is interesting to note that this period in Indian history has often been seen as a period of agrarian expansion. Western Rajasthan, especially the interiors of Thar Desert offer us glimpses of a different set of irrigation system, if not new. A closer examination of the process has been offered in the following couple of sub-sections to argue that variability in monsoon resulted in changed waterscape and perhaps culminated in a different set of irrigation systems and consolidation of settlements which have expanded its area of influence and has continued ever since.

CHANGES AND ADAPTATIONS: GROWING RELIANCE ON RAINWATER IRRIGATION

Like any other region solely dependent upon the monsoon, the climatic variability in the case of Rajasthan also is greatly associated with the timely onset and regular spread of the monsoon. It is all the more applicable for this region, that is, western Rajasthan, simply because monsoon rains are the only source of water. There is not a single perennial river in the region and ground water is available at great depth and is often brackish.[27] It has been suggested that this was not the case earlier when at least two rivers flowed through this region and sustained human settlements.[28] It has been a matter of debate to argue the relative significance of these rivers with respect to their contribution to the growth and/or decline of the urban settlements or who were the architects of these settlements.[29] To me what is more relevant is the fact that general dry climatic conditions in the region were aggravated by a gradual shift in the course of these two rivers perhaps due to tectonic movement.[30] However, the dry climatic conditions in the region during this time

[27] Nainsi, *Munhot Nainsi ri Khyat*, vol. II, p. 286.

[28] Ghosh, 'The Rajputana Desert'.

[29] Rajalakshmi, 'The Riddle of a River'.

[30] Ghosh, 'The Rajputana Desert'.

cannot be ignored and demand an investigation in the process of adaptation. Gradual non-availability of the perennial source of water is reflected in a new set of negotiations between this seasonal availability of water and the settlement patterns. Incidentally, we witness even the beginning of state formations in the interior of the Thar Desert. It appears that by this time something radical had happened which facilitated not only the continuum but growth of human settlements even in marginal regions. It appears that by this time society had understood the mechanism of monsoon-associated vagaries and was able to adapt to even an increase in aridity. This adaptation also took care of the flux associated with water which is usually more pronounced in monsoon-dependent economies. However, before moving into the thick of my arguments to account for these changes, it will not be out of place to reiterate the significance of monsoon rains in the regional economies. At the same time the implications of droughts, which were an integral part of any monsoon-dependent regions must be examined. It seems societies were able to recognize that given their dependence on monsoon rains they should develop mechanisms to mitigate the impacts of droughts.

Given the duration of long periods of arid conditions it was but natural to witness changes in the settlement patterns. Here it is important to keep in mind that Harappan, and later on during the Rangmahal culture, in northern Rajasthan agriculture was primarily riparian and was dependent on annual floods for agrarian production. B.D. Chattopadhyaya has argued that, 'the evidence seems to be more specific at Rang Mahal where in early times a high rainfall rate and annual flooding of the Ghaggar, facilitating rice cultivation . . . '.[31] However, changes in river flows must have affected annual floods and as a result agrarian production possibilities must have also been hampered.[32] Ghosh has argued that possibly these factors led to the decline of the Rangmahal culture. Decline of this culture perhaps resulted in migration after the seventh century and possibly had encouraged nomadic life.[33]

[31] Chattopadhyaya, 'Irrigation in Early Medieval Rajasthan', p. 301.

[32] Raverty, *The Mihran of Sind.*

[33] Ghosh, 'The Rajputana Desert'.

The decline of the Rangmahal culture in the beginning of the early medieval period is not contested any more, however, what followed after that in the region has generated a long debate. As pointed out earlier, on the one hand, urban settlements declined in the north and north-west Rajasthan and on the other, new and smaller settlements emerged in the interiors of the Indian desert and adjoining areas. The emergence of Bhatis in the interiors of the Indian desert in the area now identified with Jaisalmer is cited on the one hand as an example of the enhanced aridity in Bhatner region and elimination of any possibility of agriculture which forced them to migrate.[34] On the other hand, it is seen as

[34] Nevertheless, on a different note, most of the historians and archaeologists argue that the decline of Rangmahal was coterminous with the decline of water flow in the rivers Ghaggar and Hakra. (Dhavalikar, 'Green Imperialism'.) These two rivers were converted to seasonal rivers of monsoon. However, G.S.L. Devra has challenged this coupling and suggests that the decline of the Rangmahal culture on the one hand was not coterminous with the decline of water flow in the river Ghaggar and Hakra, neither did it lead to a decline of urbanization in the region. (Devra, 'Physiography and Environment'.) While accepting that the seventh and eighth century witnessed long eras of deficient rainfall, he cites evidence for the continuance of urban settlement in the region till the medieval centuries. He argues that deficient rainfall does not symbolize an extension of aridity and desert climate in the area till late in the middle ages. (Devra, 'Problems in the Delimitation'.) He has focused on north and north-west Rajasthan and extensively cites continuance of the Lakhi Jungle till at least the 13th century as evidence too suggests that the desert had not reached Bhatner and adjoining areas. (Devra, 'Desertification and Problems of Delimitation of the Rajputana Desert During the Medieval Period', in M.K. Bhasin and Veena Bhasin (eds.), *Rajasthan Ecology, Culture and Society*, Kamla-Raj Enterprises, Delhi, 1999, pp. 97–10.) He suggests that, 'Sherkhan, cousin of Sultan Balban and governor of north-west frontiers made Bhatner his headquarters. Not only he repaired the fort but also crushed the local powers of the region and at time protected the frontiers while sitting there.... All these accounts support this assumption that the region of Bhatner which was a seat of important political activities was still away from the reach of the desert.' (Devra, 'Problems in the Delimitation', p. 375.) Therefore, even if there is evidence of increased aridity, we find evidence of continuance of urban settlements.

Attributing the continuance of Bhatner as an urban centre to the agrarian production in the hinterland, Devra points out that, 'The revenue figures of the *Pargana* as they are available in the archival records of Bikaner, clearly state that it was an area of two crops in a year.' Further, while offering an explanation for

an example of colonization of new areas for settlement purposes. Similarly, 'it is perhaps significant that evidence relating to crop-production and the emergence of settlements in water scarcity areas like Marwar does not date back to a period much earlier than the early medieval'.[35] It is normal to expect that increased aridity in one region usually forced migration, but what perplexes one is the fact that newer settlements emerged in a region which was more arid and hostile for human settlements. The emergence of new settlements in arid parts of Rajasthan during the seventh to tenth century is well attested; therefore, it will be interesting to explore the processes of adaptations and innovations.

Most of the processes of *adaptation* generally relates to both technological adjustments as well as changes in human social relationships. Also, human perceptions of climatic changes are specific to individual societies and very much influence how a particular population will respond in the face of environmental shifts. This is intimately tied to how a society defines the place of humanity in nature, how climatic change is perceived in terms of its magnitude and duration and the influence of social memory on planning strategies. Different segments of society have different motivations for action when faced with environmental change. These can result in a diverse range of strategies. Social and technological resilience versus resistance to change can be critical factors in the ability of societies to adapt to new environmental conditions. Resilience may vary between segments of society leading to greater resistance to change among those members of society who benefit from current conditions.[36]

---

this feature he suggests that, 'The situation is important in itself because in most parts of the Bikaner state only one crop, i.e., *kharif* was sown in a year. Besides it, the agriculture of Bikaner state was entirely dependent upon favourable rainy seasons whereas the *Rabi* crop in the *Pargana* Bhatner had the irrigation facilities readily available from the Ghaggar.' (Devra, 'Problems in the Delimitation', p. 377.)

[35] Chattopadhyaya, 'Irrigation in Early Medieval Rajasthan', p. 309.

[36] Arlene Miller Rosen, *Social Responses to Climate Changes in the Ancient Near East*, AltaMira Press, Maryland, 2007, p. 172; For detailed discussion one can see Roderick J. McIntosh, Joseph A. Tainter and Susan Keech McIntosh (eds.), *The Way the Wind Blows: Climate, History, and Human Action*, Columbia University Press, New York, 2000.

It is therefore pertinent to examine the nature of adaptations, if any, visible in the region of changed climate and later on I will examine the factors which perhaps facilitated human settlements in the more arid regions. These are proposed here as possible factors rather than a stipulated order of events. It is more of an attempt to challenge the stereotypical 'cause and effect' relationship usually proposed for human–nature interactions. Human social response to climatic variability will vary from society to society and even within a society. While examining human adaptations to climatic variability, it is significant to bear in mind that:

> Some responses are successful and appear to be adaptive; others are less so and appear in the archaeological record as abandonments or social collapse. The choices societies make relate to their worldview, their type of social and political structures, as well as their level of technological development. . . . In most cases we may never know the motivations of such populations and why they succeeded or failed.[37]

Most visible area of adaptation which resulted initially in the growth settlements and later on state-formation during this period of climate variability can be seen in the field of irrigation and agriculture. Possibilities of the existence of irrigation systems even during the earlier period cannot be ruled out as it has been rightly argued that:

> while higher rainfall levels of the fourth millennium BC may have promoted the more widespread adoption of cultivation and sedentary settlement, subsequent decline in rainfall may have contributed to growing population density in certain regions. . . . Changing cultural practices involving some kind of water management—even at very low scale—may also have encouraged this concentration of population along water courses, although such practices are conjectural and not yet documented with archaeological populations in the river valley.[38]

Nevertheless, it is valid to argue that reduction in the rainfall levels during the early medieval times resulted in the growth and development of artificial irrigation; B.D. Chattopadhyaya, on the

[37] Rosen, *Social Responses*, p. 175.

[38] Madella and Fuller, 'Palaeoecology and the Harappan Civilisation', pp. 1297–8.

basis of epigraphic evidences, argues that 'If these instances are taken to form any generalization regarding the early historical period, then the organization of artificial irrigation in early medieval times certainly constitute a departure from the earlier pattern.'[39] This departure can be interpreted as developed in response to the changed climatic conditions and reflects one set of adaptation. As such riparian agriculture primarily depends on natural inundations and could flourish in stable climatic conditions.

Distinction between two archaic terms offers a very apt description of the region from where migration is supposed to have started and where it culminated: *nadimatrka* (river fed) or *devamatrka* (rain-fed).[40] Technically these two terms describe two different types of agricultural regions based on irrigation methods and more or less encapsulate the adaptation to climatic variability and changed waterscape. The interiors of the Indian desert had witnessed the growth of some very ingenious methods of irrigation which made agriculture possible in this water-deficient region. Traditionally these irrigation devices have been clubbed into two categories. 'We may start with the rather obvious statement that artificial irrigation in early medieval Rajasthan was provided by (1) Tanks and (2) Wells.'[41] However, these two broad categories fail to capture the totality of human adaptations in this arid region. There were several other methods of irrigation which utilized rainwater and harvested it for agricultural purposes like *khareen*, *rela*, *sewanj*, *bahla*, etc. These systems basically stored rainwater to be used for irrigation.[42]

The most unique system among them had been practised in Jaisalmer region and was known as *khareen*.[43] It is generally believed that Paliwal Brahmins who settled in Jaisalmer region during the early medieval times developed *khareens*. The land

[39] Chattopadhyaya, 'Irrigation in Early Medieval Rajasthan', p. 301.

[40] Ibid., p. 298.

[41] Ibid., p. 301.

[42] Kumar, 'Situating the Environment'.

[43] Nainsi, *Marwar-ra-Pargana-ri-Vigat*, vol. II, p. 12 and Chand, *Tawarikh Jaisalmer*, p. 92.

irrigated through *khareens* supported a rich harvest of wheat.[44] *Khareens* were basically *bunds* created in the path of *nullahs* (seasonal streams).[45] The height of the *bund* was kept low so as to retain only limited water, which could dry up in two to three months. This dried tank bed was used to produce *rabi/unali* crops. Another advantage of keeping the height of the *bund* low was that the overflow water could be utilized to create another *khareen* downstream, which would in turn provide a larger cultivable land. It is important to note that cultivation with the help of *khareen* could be achieved in the region adjoining the hills where soil was largely clay and it was irrelevant for the sandy soil.

It is interesting to note that the irrigation techniques they developed in the region resemble the farming systems of the Mediterranean world developed between fourth century BC and seventh century AD.[46] It is difficult to map the route through which these techniques reached Rajasthan; or to determine if these had indigenous origins. However, it is interesting to note that Baluchistan (now in Pakistan) offers several evidences of the deployment of a similar technique to manage water in this arid land and is locally known as Gabar Bund.

> Archaeological investigation shows that human beings living there (Sindh Kohistan region located between the main hilly ranges of Baluchistan and the Indus alluvial plains) have learnt to utilize the rainwater efficiently and have established excellent water storage system known as "Gabar Bund", where rainwater is collected artificially for agriculture and other usage.[47]

[44] Mishra, *Rajasthan ki Rajat Boonden*, p. 62.

[45] S. Ali Nadeem Rezavi, 'Kuldhara in Jaisalmer State: Social and Economic Implications of the Remains of Medieval Settlement,' *Proceedings of Indian History Congress*, 56th Session, Rabindra Bharti, Calcutta, 1995, pp. 312-38.

[46] Hanoch Lavee, Jean Poesen and Aaron Yair, 'Evidence of High Efficiency Water-Harvesting by Ancient Farmers in the Negev Desert, Israel', *Journal of Arid Environments*, vol. XXXV, no. 2, February 1997, pp. 341–8.

[47] Qasid Hussain Mallah, 'An Archaeological Assessment of Taung Valley of Sindh-Kohistan Pakistan', *Ancient Asia*, vol. II, 2009-10, pp. 93–112.

Mallah, on the basis of extensive availability of artefacts of the Chalcolithic Nal Period, on the very sites of Gabar Bund has described this system as indigenous to this region:

> The hydraulic strategy or the water management system adopted by the locals in this region during the pre-modern time is also very interesting as it would have created availability of water throughout the year even in arid and semiarid conditions. A network of stone walls was built at strategic points to strategically manoeuvre the flow of monsoon rivulets into the mainstream and was used to flood the alluvial valley. Consequently an agricultural system based on, Barani (rainfed) crops were possible and also the water could be stored in several spots within the bed of Nai for consumption throughout the year. Though it is very difficult to date these Gabarbands and the stone walls they belong to the pre-modern times probably even to the prehistoric Harappan civilization as water management systems was a common practice utilized by them for survival in the arid zone. . . . The Gabarband system in Baluchistan has its beginning in the Chalcolithic Nal period. Based on the term "Gabar", historians have associated this system to the "Medieval Iranian Zartushtis who are also called Parsees". . . . These Zartushtis were fire worshippers and mainly depended upon agriculture besides herding. However, the discovery of prehistoric sites located in close vicinity help substantiate the primary notion of archaeologists regarding very early their origin in the area.[48]

Thus suggesting round-the-year availability of water; during droughts, people would cluster around such perennial sources of water.

The sequence of transmission from the Mediterranean to the Thar receives further credibility once we consider evidences for similar systems in Ancient Iran. *Encyclopaedia Iranica* suggests that:

> A number of massive stone dams, now known in the archaeological literature as *gabar-band*s, appear to have supported terraced fields in the hills bordering the main depression. . . . This type of engineering continues to be practiced on a small scale throughout Baluchistan (and in other parts of Afghanistan, Iran, and Pakistan; . . .).

[48] Ibid., p. 93.

It is significant to note that this system was primarily practiced in the region known as Kharan.[49] A possible correlation between the term, *Khareen*, used for a similar system and the region of Kharan cannot be entirely negated. This in most probability points towards the foreign origin of the system.[50]

None the less, it appears that these systems were modified according to the climate and the landscape of Jaisalmer and adjoining areas.

In the pan-Mediterranean area agricultural intensification in hilly areas took the form of terrace systems that were constructed horizontally along the slopes, parallel to the valleys. This system worked well to preserve the deep soils of the slopes and capture water in a series of stepped fields.[51]

Whereas,

In the Negev region farmers adapted this typically Mediterranean system by changing the orientation of the terracing to a vertical configuration vis-à-vis the slopes, allowing water to run off from the hill slopes more effectively and to be concentrated on the floodplains of the *wadis*, held in place by a series of horizontal check dams across the *wadi*, at the right angles to the flow. This type of water trap was also a sediment trap, preserving fine-grained material in the channel and renewing the fields with both fresh silt and moisture.[52]

On the basis of the field experimentation, it has been demonstrated that this agricultural technology was used to produce much higher yields of a number of different crops than would otherwise have been possible in an area with such low rainfall amounts. This allowed Negev towns to support large populations and enter into the Mediterranean agricultural production economy by producing wine for export as well as internal consumption.[53]

[49] Brian Spooner, '"Baluchistan", Geography, History and Ethnography', in Ehsan Yarshater (Gen. ed.), *The Encyclopedia Iranica*, Center for Iranian Studies, Columbia University, New York, http://www.iranicaonline.org/articles/baluchistan-i (accessed on 18 September 2011).

[50] I owe this information to Prof. Pankaj Tandon.

[51] Rosen, *Social Responses*, p. 162.

[52] Ibid.

[53] Ibid., p. 164; records of yields at the Byzantine period site of Netzana have come down to us from a cache of papyri uncovered at the site. These textual records inform us that wheat, barley, grapes, olives, figs and dates were cultivated

The potential of this technique can be appreciated if we consider the fact that, 'Jaisalmer district in Rajasthan receives an annual rainfall around 4 inches and therefore can collect around 2.5 gallons water per square foot of catchment area.'[54] Thus, it can be safely argued that there was potential in rainwater harvesting in the arid parts of Rajasthan. It is important to note the significance of topographical and geological factors in the success of this technology. Sheer availability of water was not sufficient to ensure agriculture in a landscape dominated by sandy soil. There are still vast expanses of sandy soil in the neighbouring area with similar amounts of rainfall, but here one cannot practice agriculture due to the nature of the soil. Most of these *khareens* were located on the slopes of occasional hillocks of the region. These *khareens* were constructed on the path of drainage which carried rainwater from the hills. Such a location enhanced the quantity of water available at the *khareen*, thereby maximizing the gains from the meagre rainfall. Another very significant factor which contributed to the possibilities of agriculture was the silt carried with the drainage. Silt was generated by the runoff along the hillocks and carried sediments to change the composition of the soil and made it feasible for agriculture. Sand was gradually converted to sandy loam along the path of the runoff and this feature was exploited to sustain agriculture.

It will not be inappropriate here to mention that Paliwal Brahmins, who are attributed with the introduction of this technology in this region, were original inhabitants of a region which is located along the foothills of the Aravalis.[55] Therefore, it will not be wrong to postulate that Paliwal Brahmins were aware of the potential offered by the runoff of the hillocks in their original place of habitation. Moreover, given the higher amount of rainfall

---

by the local population in the vicinity of the town. Wheat was by far the most important crop and produced yields averaging close to sevenfold that of the initial amount of seed grain planted, or in the neighbourhood of 6,000 bushels.

[54] Pooja Nagrath, 'A Comparative Analysis of Rainwater Harvesting Practices Implemented in Arizona and Rajasthan', unpublished M.Sc. thesis, Arizona State University, Tempe, 2008, p. 26.

[55] Rezavi, 'Kuldhara in Jaisalmer State'.

in their region there was no need to construct *khareens*. Therefore, it is not surprising to note that initial habitation in this region has been traced to Pali and similar neighbouring areas with higher rainfall.[56] To me it appears that B.D. Chattopadhyaya could not appreciate the topography of the region and has inadvertently equated *nalabua na lopya*[57] with well irrigation.

Here, it is pertinent to point out that the well-documented agrarian expansion of early middle ages in other parts of the Indian peninsula should not be seen only as a by-product of urban decline as postulated by R.S. Sharma.[58] On the other hand, the agrarian expansion and growth of settlements have been explained in terms of developments in irrigation and resultant shifts and changes. It has been aptly deciphered by B.D. Chattopadhyaya on the basis of epigraphic evidences that 'the organization of artificial irrigation in early medieval times certainly constitute a departure from the earlier pattern'.[59] As in the case of some areas of Rajasthan, the expansion of the agrarian base and rural settlement region in Tamil Nadu too was linked with the expansion of the irrigation network. R. Champakalakshmi has suggested that 'there is a general correspondence between the steady increase in irrigation works and the increase in the nadus under the Pallavas and early Cholas'.[60] Similar changes in the nature of irrigation and the resultant social edifice has been documented for other regions as well. 'An artificial irrigation was the most important factor for the agriculture in Telingana, construction of tanks was essential to attract settlements. . . .'[61] Growing reliance on artificial irrigation has also been documented for the south Karnataka region.[62] This

56 Chattopadhyaya, 'Irrigation in Early Medieval Rajasthan', p. 311.

57 Chattopadhyaya, 'Origins of the Rajputs.'

58 Sharma, *Urban Decay in India*.

59 Chattopadhyaya, 'Irrigation in Early Medieval Rajasthan', p. 301.

60 R. Champakalakshmi, 'The Study of Settlement Patterns in the Cola Period: Some Perspectives', *Man and Environment*, vol. XIV, 1989, p. 92.

61 I. Lakshmi, 'Wet Crop Cultivation, the Brahamans and the Cultivating Communities in Medieval Telingana', *Indian Historical Review*, vol. XXXI, nos. 1 & 2, Jan–July 2004, p. 61.

62 Chattopadhyaya, *Aspects of Rural Settlements*.

departure can be interpreted as developed in response to the changed climatic conditions and reflects one set of adaptation.[63]

The referral of Paliwal Brahmins as the custodian of the new irrigation system, that is, *khareens*, suggests that the colonization of the newer region was attained under the guidance and patronage of the Brahmins. Although the indigenous genesis of the new irrigation system, *khareens* is yet to be established, none the less, the prominent role of the Brahmins cannot be ignored. 'Agrarian expansion was boosted by the migration of brahmanas and craftsmen who disseminated advanced knowledge of agriculture and technology.'[64] Furthermore, it must not be ignored that along with the Brahmins, it was political contingency on the part of the ruling dynasties to promote Brahminic domination to strengthen the legitimacy of the dynasty.

> Now the question is, why were the brahamanas preferred? The first reason is their knowledge of wet rice cultivation, [in our case knowledge of Khareens]. . . . Apart from the above-stated reasons the newly settled regions had to be structures to maintain order and brahmanas from regions of earlier kingdoms could serve as agents to introduce the desired structure and culture.[65]

The migration of Brahmins from the towns and urban centres to enjoy land grants has been very painstakingly documented by R.S. Sharma for the Gujarat region.[66] At the level of conjuncture, I will not hesitate to suggest that the migration of Paliwal Brahmins to the Jaisalmer region does not seem improbable; rather it is representatives of the trend visible in other parts of the Indian peninsula. Although need for further research to clearly establish this tendency cannot be ruled out.

[63] B.D. Chattopadhyaya, 'Early Memorial Stones of Rajasthan: A Preliminary Analysis of their Inscriptions,' in B.D. Chattopadhyaya, *The Making of Early Medieval India*, Oxford University Press, Delhi, 1994, pp. 120–9.

[64] Sharma, *Economic History of Early India*, p. 150.

[65] Lakshmi, 'Wet Crop Cultivation,' p. 68.

[66] Appendix I, 'Towns in Gujarat/Malwa/Karnataka from where the Brahmans Migrated to Enjoy Land Grants in Gujarat', Sharma, *Urban Decay in India*, pp. 186–90.

Moreover, this system of irrigation thus supposedly developed by the Paliwal Brahmins was well adapted for this region as these systems not only provided water for irrigation but also helped in ensuring the productivity of the land by accumulating sediments carried from the hills. It must have also raised the water level which could have been used to secure potable water through wells. Application of a similar principal has been documented for a later period for the same region by the author:

Another important method to harness the rains was *bera*. These were constructed near the source of water or on the dry bed of a water body. The technique used in the construction of *bera* was somewhat similar to the ones used in the construction of *kuin/teen*. The percolation of water from any water-body to ground around it provided an opportunity to tap it by construction of *beri*, a kind of well. This technique was quite prevalent especially in western Rajasthan. The village Godhali of *pargana* Phalodi had *beri,* which received water from nearby seasonal streams. Similarly, village Naberi of *pargana* Phalodi could harness water in the dry bed of seasonal stream. The village Bhabhclai of *pargana* Jodhpur had no source for 'drinking water', underground water being salty/ brackish. The only source of water was *beri* sustaining on a nearby tank. Similarly, village Mahev in *pargana* Sojhat had *bera* in a *talab,* which could retain water only for 8 to 10 months. Village Chanda Vasani also had a *bera*. In Jaisalmer the bed of Amarsagar has several *beris*. The tank tends to retain water for only six to eight months and for the remaining period these *beris* supply water for drinking purposes.

The above-described technique was used in those parts of Rajasthan where ground water was very deep and retention of water in the ponds was possible only for a few months. In such a situation, people used to dig wells on the bed of dried pond. Water could be retrieved because of its percolation in the bed.[67]

Moreover, epigraphic evidences very clearly point towards the growth of water management systems. 'Two epigraphs recorded in the Vikrama samvat (viz. VS 1053/AD 997 and VS 1097/AD 1040),

[67] Kumar, 'Flexibility and Adaptability', pp. 41–2; B.L Bhadani and Jibraeil, 'Technique of Irrigation in the Desert area of Phalodi in Medieval times,' *Proceedings of Rajasthan History Congress*, vol. XXIII, Bikaner Session, 2008, pp. 36-43.

located at Dabala (Some nine miles south-east of Jaisalmer town) mention the foundation of a tank and well by the "mahidhara" (king) vatsya, son of the Sodha Jajjila . . .'[68] Similarly, 'An epigraph of July AD 1031 . . . records the establishment of a tank by Rinapara, the servant of one Raja sri Kambhasimha, near Budha Dumgar'.[69] Furthermore, one can still locate wells, step wells close to the water bodies especially on the periphery of the *khareens* in the interior parts of Thar Desert. One such step well has been excavated near a *khareen* in Kuldhara, the ancient capital of the Bhatis some 25 km from Jaisalmer.

In the interior of Thar Desert one can see the prevalence of another unique system to harness rainwater as pointed out by the author:

> One such method of water harvesting was known as *kuin* or *par* or *teen*. It was quite similar to a well in appearance, however, it was worked on a different principle. Due to the presence of an underground layer of chalk or limestone, which restricted percolated rainwater from mingling with the brackish ground water, there was a possibility to extract the tapped water. The presence of the underground layer of limestone ensured that rainwater would remain potable and could be appropriated. Moreover, sand does not bind up like soil, hence restricting the possibilities of evaporation of trapped percolated water. Rainwater so retained by sand gradually seeps into the open space available at the bed of *kuin*. It was usually not more than three to four earthen pots in a day. Therefore, *kuin* was not meant for round the clock water availability and was used for drinking purposes only. The sheer fact that prevalence of such wells is reported by Nainsi suggests the significance of such devices as it ensured continuance of human settlement. This technique speaks for the better appropriation of given environmental conditions and better management of limited natural resources.[70]

It is relevant here to point out that most of the *beris* constructed on the river beds were not brick-lined as it would have restricted percolation of water into it during the lean season and would have

[68] A.G. O'Brien, *The Ancient Chronology of Thar: The Bhattika, Laulika and Sindh Eras*, Oxford University Press, Delhi, 1996, p. 51.

[69] Ibid., pp. 51–2.

[70] Kumar, 'Ecology and Traditional', pp. 74–5.

impinged on the fertile agricultural land as well. In water-deficient regions every inch of land with moisture suitable for agriculture must have been precious. However, it is possible that because these were not brick-lined, we do not get any archaeological evidences for that. Apart from the fact that the construction of a brick-lined well on the bed would have disturbed the plain of the bed used for agricultural purposes, moreover, it was cost intensive as well. Even during the medieval times when state formation had reached its 'zenith' and the state was actively involved in the protection of settlements within their domain the proportion of brick-lined wells was very small compared to earthen wells.[71] Similarly, even the *kuins* were earthen in nature and at the most were supported by rope, therefore the possibility of survival of archaeological evidences is very bleak. Future archaeological excavations might offer us some clue. Apparently, geo-physical conditions of the region suggest limited or negligible possibilities of human settlement; however, the prevalence of such methods of water harvesting clearly suggests possibilities of human settlements in the region.

The statement, 'Evidence on crop production in early medieval period has also to be strenuously culled from the mostly indirect information that the inscriptions offer',[72] captures the limitations any historian working on such issues faces and also explains the reasons why he is forced to conjecture on the basis of the divergent facts. Prevalence of *khareens* in the arid part of Rajasthan created possibilities of the cultivation of *rabi* (winter) crops in general and wheat in particular. On the basis of epigraphic evidences for the period B.D. Chattopadhyaya argued that:

> An inscription of 1059 from Panahera (Banswara) may refer to rice-field irrigated by *arghattas*. Cultivation of *gudhuma* (wheat) appears to have been on a larger scale and is attested by a number of inscriptions. . . . This leaves some room for postulating a connection between territorial expansion of agriculture and artificial irrigation.[73]

Moreover, it is significant to note that, 'The law books lay down

[71] Kumar, 'Situating the Environment'.

[72] Chattopadhyaya, 'Irrigation in Early Medieval Rajasthan', p. 308.

[73] Ibid., p. 309.

severe punishments for those who cause damage to tanks, wells, ponds and embankments, etc.'[74]

Thus, the prevalence of *khareen* and other methods of rainwater harvesting[75] might have offered the required impetus for human settlement in the region. Moreover, any increase in winter precipitation as postulated for an earlier period by Modella and Fuller[76] would have further enhanced the productivity or even if there was none, accumulation of the monsoon runoff would have ensured sufficient moisture. The role of technology has been stressed in the gradual settlement of uninhabited and resettlement of already inhabited or once settled but now deserted areas.[77] Gradual expansion of irrigation facilities and resultant expansion and improvement of agriculture suggest greater utilization of natural resources. Subsequent emergence of regional kingdoms and their expansion could not have taken place in the absence of agricultural surplus generated largely due to better irrigation facilities. Nevertheless, it will be inappropriate to believe that settlements in the marginal areas could have emerged solely on the basis of agriculture. Historically it has been demonstrated that the beginning of agriculture in any region has mostly been supported by the cattle-rearing traditions.[78]

[74] Sharma, *Urban Decay in India*, p. 173.

[75] Kumar, 'Ecology and Traditional'; the prevalence of irrigation-based agriculture during this period in the region is well documented. 'Apart from the inscriptions of the Guhilas of Kishkindha and Dhanagarta, which refer to irrigation-based agriculture, a more specific connection is suggested by a few records of the Mandor Pratiharas', Chattopadhyaya, 'Origins of the Rajputs', pp. 62–3.

[76] Madella and Fuller, 'Palaeoecology and the Harappan', p. 1297.

[77] Chattopadhyaya, 'Irrigation in Early Medieval Rajasthan', p. 51.

[78] Brian M. Fagan, *People of the Earth: An Introduction to World Prehistory*, Scott, Foresman & Company, Glenview, Illinois, Boston, London, 1989, pp. 251–374; Robert J. Venke, *Patterns in Prehistory: Humankind's First Three Million Years*, Oxford University Press, New York & Oxford, 4th edn., 1999, pp. 268–330 and 484–513.

CHANGES AND ADAPTATIONS: 'AGRO-PASTORALISM'

It will be inappropriate to explain the emergence and growth of habitation in agriculturally marginal areas solely in terms of adaptation to a different regime of irrigation practices and resultant agrarian production. As such it has been historically well argued that in the beginning agriculture cannot be the sole source of sustenance. It needs to be complemented by supplementary sources of sustenance and more often than not, cattle-rearing has been extensively relied upon along with long-distance trade. Therefore it is not surprising to note that most of the settlements which emerged during this period practiced a combination of different modes of sustenance. Most of these societies combined agriculture with animal husbandry and the trend continued till very recent times. It has been time and again suggested that the region of Jaisalmer was colonized by the people who had a pastoralist background.[79] It is interesting to note that Madella and Fuller, arguing for an earlier period, suggest that, 'it is also probably during the later fourth to early third millennium cal BC that pastoralism became more widespread in Gujarat and Rajasthan. It is plausible that the wetter mid Holocene created more potential grazing area.'[80] It is not surprising to note that two diagonally opposite waterscapes have been described as a cause for the emergence of pastoralism in the region. Nevertheless, it can be safely argued that in this region agriculture and pastoralism coexisted and may have possibly helped local communities mitigate climatic variability. Moreover, it was during the early medieval times that we get evidence for the introduction of camels in the region.[81]

Gradual movement from the well-developed 'urban culture of Rangmahal' to the sporadic settlements in the interiors of the desert though at a lesser level reflects an excellent process of adaptation

[79] O'Brien, *The Ancient Chronology of Thar*; Tod, *Annals and Antiquities*, vol. I.

[80] Madella and Fuller, 'Palaeoecology and the Harappan Civilisation', p. 1297.

[81] S.P. Gupta, 'Reconstructing the Political and Economic Profile of Rajasthan,' *Proceedings of Indian History Congress*, 55th Session, Aligarh, 1994, p. 61.

and it has largely been ignored by historians and archaeologists. Survival reflects a process somewhere between the experience of the Maya and Chimu civilizations;[82] where urban centres faltered in the region, rural settlements at the smaller level were able to survive. It was perhaps at the rural level that they were able to diversify their resource base/s both in terms of spatial spread as well as sustenance patterns. At another level they were also able to integrate themselves with long-distance trade which provided them with the necessary buffer to mitigate really harsh drought periods. Pastoral practices gave them flexibility to exploit the potential offered by seasonal monsoon rains by growing agriculture and/or ultimately temporary migration towards wetter areas or dispersing to reduce the concentration of burden on a given landscape. Mobility offered them possibilities to harness resources of neighbouring areas and seasonal agriculture offered them cereals to be consumed later on. Unlike the hunting gathering stage where the seasonal availability of vegetation conditioned movement, agriculture offered them the liberty to plan their migration with more flexibility. At the same time, it is pertinent to note that true pastoralism is difficult to sustain and it has been able to sustain only with the support of an agrarian economy,[83] thus, we witness a combination of both modes of sustenance.

Historically, evidences of 'hero-stones' in this region for this period[84] are an important testimony to argue for the growing premium on the pastoralism. 'This also suggests that cattle keeping had become more widespread than agriculture. Pastoralism had become a mode of subsistence, and hence frequent cattle raids.'[85] Another important facet of hero stones to be carefully

[82] Fagan, *The Long Summer*; *The Great Warming*.

[83] Anatoly M. Khazanov, *Nomads and the Outside World*, University of Wisconsin Press, Madison, 1994, pp. 67 and 191; Peter B. Golden, *An Introduction to the History of the Turkic Peoples: Ethnogenesis and State Formation in Medieval and Early Modern Eurasia and the Middle East*, Otto Harrassowitz, Wiesbarden, 1992, p. 8.

[84] Chattopadhyaya, 'Early Memorial Stones of Rajasthan', p. 128.

[85] The evidence of the hero stones indicates that cattle raids had become a regular feature in the post-Gupta times, and protecting cattle from raiders had

considered is the nomenclature. 'The pillars are locally known as *govardhanas* and possibly also as *tirthambs*. The term *govardhana* or *govardhandvaja* is as early as the memorials themselves.'[86] The literal meaning of the term *govardhan* means a place and/or thing which promote growth of cattle.[87] The significance of drought animals is also evident from the numerous references to cattle-loot in the medieval literature.[88] Cattle were considered important economic assets and were offered as part of dowry.[89] Traditional feuds were also settled by cattle loot.[90]

Perhaps due to a very different ecological setting of Jaisalmer region, the terms *govardhans* attains a unique meaning. A meaning which is apparently quite different from the conventional one but at the same time broadly resonate with similar socio-ecological function. These are independent monolithic structures which usually also carries inscriptions. The top of these stone pillars is usually adorned with the sculptures of gods and goddesses. Although there are few *govardhans* which carry the sculpture of a hero usually on a horse, but locals claim it to be one of the *panch-pirs*. The main beam of the pillar carries inscriptions. These were generally placed along with water bodies to commemorate its construction. It is very clear from the inscriptions that most of these *govardhans* were erected to eulogize the acts of valour and religious

---

become the main concern. Village youths often encountered the cattle lifters and in the ensuing skirmish sometimes someone among them was killed. It is in the memory of such people that a hero stone was erected. One can also trace the coexistence of pastoral communities along agriculture in similar arid climatic conditions of Maharashtra. See Gunther-Dietz Sontheimer, *Pastoral Deities in Western India*, tr. Anne Feldhaus, Oxford University Press, New York, 1989. However, rivers in this region enhanced possibilities of agriculture and contest over space can be seen in the religious practices as described in the river *mahatmyas*. See Anne Feldhaus, *Water and Womanhood: Religious Meanings of Rivers in Maharashtra*, Oxford University Press, New York, 1995.

86 Chattopadhyaya, 'Early Memorial Stones of Rajasthan', p. 121.

87 It is a combination of two terms, *go* (cow) and *vardhana* (increase).

88 Nainsi, *Munhot Nainsi ri Khyat*, vol. II, p. 128 and vol. 3, p. 202.

89 Ibid., vol. III, p. 63. Pabuji promised his sister to give camels in dowry.

90 Ibid., vol. II, p. 144 and vol. III, p. 15.

merit.[91] The present level of historical investigations does not allow to conclusively define the primary function of these water bodies; whether for irrigational or potable purposes? It is interesting to note that more often than not mercantile communities were donators for the construction of these water-bodies. Moreover, historically we can trace that this region has traditionally been patronized by mercantile community. Nevertheless, on the basis of very limited survey of the region, I will argue that the primary function of these water-bodies was to ensure availability of water for potable purposes. Their spatial distribution also supports this line of argument. Furthermore, it will be relevant to suggest that these widely distributed water-bodies supposedly supported pastoral communities and must have also been a great help to the mercantile community.

The official records and popular literature are full of evidence symbolizing the importance of cattle wealth in the society. The important animals were goat, sheep, camels, bullocks, cows, horses, etc. It will be pertinent on our part to elucidate the strategies in response to the regional environmental constraints which often manifested in different strategies of adaptations for different sections of the society. This was visible in the variations in the herding strategy in terms of the numbers and species of animals reared and combinations of animals in different locations. Most of the animals employed by these groups were energy conservants. For instance, the capacity of the camel to withstand shortage and irregularity of water supply (which was the most limiting factor here) for a number of days (without having an effect on lactation) was especially crucial. Its long gestation period, however, made it more useful for milk, while the fairly rapid build-up of herds, which the short gestation period of sheep and goats, had made them indispensable as animals of slaughter, thus offering both meat and wool.

[91] Inscription of Vijayasar (eighth to tenth-eleventh centuries) As cited in Virendra Singh, *Jaisalmer Rajgharane ka Kendriya Satta se Sambandh*, Rajasthani Granthagar, Jodhpur, 2011, pp. 165-73.

In the arid region where vegetation was very meagre, sheep, goats and camels could easily survive on the wild grass like *phog*, *ken*, *bharut*, *babool*, *khejri*, etc. This is testified by the following saying:

काल कुसम्मै ना मरै, बामण बकरी उंट ।
वो मांगै को फिरै चरै, जो सूखा चाणै ढूंठ ।।[92]

It is said that the brahmin, goat and camel shall survive the period of scarcity as brahmin can survive by begging, goats can survive on wild grass and camel can eat dry trees.

The cattle/herds required pastures, the natural vegetation for most part of the arid region. Yet, we come across the social practice of keeping common land accessible for animals near the village. This had the state's approval as well. This was known as *jor*.[93]

For some reason the role of goats in the early medieval societies in Rajasthan has been generally ignored. A plausible reason can be seen in the arguments of Siddle who holds the view that,

In the official mind and in popular consciousness the goat has been the *bête noire* of dry farming areas. It has been damned as one cause of the encroachment of desert margins in Africa, the degradation of Mediterranean and Middle-Eastern hillsides and the general decline of fertility. Nor is this aspect of their reputation newly acquired—indeed until very recently goats seem to have almost always had a bad press from the authorities.[94]

That does not mean that the goat was not an integral part of the pastoral communities in this region in the ancient past.

The goat was the earliest ruminant to be domesticated. The Harappa toys contain representations of a goat. Two seals from Mohanjo-daro show a wild bezoar goat with enormous curled horns, and a bearded domestic male goat with side-spreading horns. The Gaddi goat, which greatly

[92] Govind Agarwal, 'Rajasthan ke Jan-Jeevan aur Lok Sahitya main Oont', *Maru Bharti*, vol. XIII, 1965–66, p. 20.

[93] Nainsi, *Marwar-ra-Pargana-ri-Vigat*, vol. I, pp. 448, 452, 457, 461, 514, 516; vol. II, pp. 328 and 342.

[94] David Siddle, 'Goats, Marginality and the "Dangerous Other"', *Environment and History*, vol. XV, no. 4, November, 2009, p. 522.

resembles the ancestral wild goat, was used as a beast of burden in the mountains and is still used in the Himalayan region of India for carrying salt and food grains.[95]

The significance of the goat in this region can also be gauged by the fact that the nomenclature of Ajmer is often traced to 'Goat' since *aj* in Sanskrit means goat.[96]

Another very important factor to be considered is the ability of cows, goats and camels to survive in arid conditions. 'Cattle are efficient eaters in the sense that they excrete as little as 25 per cent of the protein they consume. This means that they can eat dried-out, low protein grass and still survive.'[97] The same can be said for goats:

> Goats represented a source not only of meat, milk and high protein cheeses but also a wide range of other uses. These varied from purely practical such as safe transport of liquids (water, wine, oil) in their skins to the recreational and ceremonial uses of skins and horns for bagpipes and for drum and horns. For a subsistence pastoral economy in a dry upland environment, the food-tolerant goat may be seen as much more valuable than sheep and certainly than the demanding bovines. Indeed, as a recent NGO campaign testified, a goat represents the best possible investment for a poor family. For a very small outlay a goat could yield between fifteen and thirty percent more value than a sheep. With no break in fertility after gestation, one pair can produce a hundred offspring in five years. . . . Moreover, goats are much better adapted to eating the woody forage of the margins of the used spaces of a community and need less to sustain life than either sheep or cattle.[98]

The case of the camel is self-explanatory as the camel seems to be designed for arid conditions and it has been suggested that precisely during the period of our study the camel was introduced in Rajasthan. In the process of growing settlements in Rajasthan, the role of the camel cannot be undermined. S.P. Gupta has

[95] R.M. Acharya, *Sheep and Goat Breeds in India*, FAO Animal Production and Health Paper 30, FAO, Rome, 1982. http://www.fao.org/docrep/004/x6532e/X6532E01.htm (accessed 17 May 2010).

[96] Sharma, *Rajasthan through the Ages*, p. 33.

[97] Fagan, *The Great Warming*, p. 55.

[98] Siddle, 'Goats, Marginality', p. 524.

suggested that the introduction of 'dromedaries, or camels with single humps . . . could have given a new economic profile to the Thar just on the eve of the middle-ages'.[99] The role of the camel has generally been relegated to the obvious and not enough attention has been paid to the significance of this animal. It is not surprising to note that Richard W. Bulliet was puzzled to note that from Morocco to Afghanistan, 'Why did the camel replace the wheel at the particular point in history when it did, that is to say, apparently after the third and before the seventh century AD?'[100] Examining the reasons for his hypothesis, he points out that as late as the later part of the nineteenth century the imperial power of Britain was seriously evaluating the feasibility of the camel over other modes of transportation in the arid landscape of Africa. Commenting on the significance of the camel, which holds good for early medieval Rajasthan as well, he writes that:

> No similar comparison of a camel with Roman ox is available, but Major Leonord did put in writing his estimation of the relative advantages of a pack camel and an oxcart in a proposal for the introduction of the former into South Africa. His conclusion is as follows:
>
> And now in a few words let us sum up the special advantage of camel over ox transport.
>
> 1. Can carry or draw twice as much. 2. Faster, and able to cover more ground daily. 3. Can do from 20 to 25 miles in one stretch. 4. Will make many more journeys in a year and in their respective lifetimes. 5. Able to traverse ground that a wagon will stick in. 6. No trouble fording rivers, where wagons would have to be unloaded. 7. [Not germane]. 8. Live and work four times as long. 9. Greater powers of abstinence from food and water. 10. Greater tenacity and endurance. 11. Wagon liable to break down, upset, or stick. Consequent loss of time and additional expense in former case, besides. 12. Lastly, the additional dead weight of the wagon which is considerable – at least a ton, I should say.[101]

[99] Gupta, 'Reconstructing the Political and Economic Profile', p. 161.

[100] Richard W. Bulliet, *The Camel and the Wheel*, Harvard University Press, Massachusetts, 1975, p. 21.

[101] Ibid., p. 23.

Thus it is not surprising to note that the camel and his activities have become similes for numerous popular sayings. It is a hardy animal and can survive in very adverse situations as stated earlier:

काल कुसम्मै न मरै, बामण बकरी उंट ।
वो मागै वो फिरै चरै, को सूखा चाणै ढूंठ ।।[102]

There have been comparisons between the camel and the goat. It is beautifully represented in the following saying which says that in the presence of camels, goats are like silver in presence of gold.

ये उंट सारै लाय टिया बाकरा, केसर सारै भाग
ये चांदी सारै लाय टिया बांबा, सोनो सारै लावो टांग [103]

According to local traditions, the camel has been placed above the goat. Camels must have had greater usefulness in the desert conditions of the western part of Rajasthan. It was given as a gift by wealthy patrons to poets, *charans*, etc. The following saying testifies this tradition,

ऐसा था पदमां कंवर, करने का लटका
हिक हिक कवि को, समपिया, कटहला कटका[104]

Poet says that the camel presented to him is of high quality. It is like a lotus and is very beautiful.

There are numerous references of the looting of camels. In the medieval period, it was a major cause of dispute. Nainsi, in his *Khyat*, records several fatal fights for the possession of camels.[105] The economic value and utility of the camel is beautifully highlighted in the following saying. It also highlights the fact that camels were so precious that they were conserved and never killed:

साड़ दागे जाते है मारे नही[106]

It means that camels were not killed even in fights rather they are captured and branded with your own symbol.

[102] Agarwal, 'Rajasthan ke Jan-Jeevan aur Lok Sahitya mein Oont', p. 20.
[103] Ibid., p. 22.
[104] Ibid., p. 23.
[105] Nainsi, *Munhot Nainsi ri Khyat*, vol. II, pp. 94, 159 and 175.
[106] Ibid., p. 223.

All three animals were useful in more than one way. Apart from milk and meat all of them must have contributed in the form of manure. We have evidence that peasants invited cattle and camel herders to put their animals on their farms.[107] If need be cows and camels were capable of draught power and should have further enhanced the capacity of the peasant to cultivate larger areas, if the waterscape permitted which was defined and redefined every year depending upon the amount, durations and timing of rains received during the monsoon. All these factors should have contributed in the initial settlement in the region even during arid conditions. Moreover, camels must have facilitated the exchange of material between settlements.[108]

Consequently, one can see the prevalence of various ethics and rituals and social-religious practices where the protection of livestock was envisaged. The significance of livestock can be gauged from the fact that the five famous saints (*panj-pir*) of Rajasthan were related to the protection of livestock. Saint Gogaji died while rescuing cows.[109] Similarly, Pabuji 'is worshipped by his rural devotees as a saviour of mankind and pastoral wealth'.[110] According to one tradition, the profounder of a local sect, Pabuji is identified as a protector of camels. It is said that Pabuji not only brought the 'reddish-brown she-camel to Marwar, but also assigned this

[107] Though studied for a later period, the significance of cow dung in agrarian economy can be gauged by the following example. Dispute over the use of cow dung was also settled by the state. It is important to note that cow dung was treated as a resource and its collection from the common land of the village was allowed to a particular caste only. *Sanad Parwana Bahi*, Asoj Vadi 2, 1822, vs, Jodhpur Records, Rajasthan State Archives, Bikaner. Henceforth, Jodhpur Records, RSAB. An examination of conflict and contest over natural resources in this region can be seen in the following works: Purnendu Kavoori, Pastoralism in Expansion: *The Transhuming Herders of Western Rajasthan*, Oxford University Press, Delhi, 1999; Mayank Kumar, 'Claims on Natural Resources'.

[108] Gupta, 'Reconstructing the Political and Economic Profile', p. 161; Bulliet, *The Camel and the Wheel*.

[109] Pemaram, *The Religious Movements in Medieval Rajasthan*, Archana Prakashan, Ajmer, 1977, p. 33.

[110] G.N. Sharma, *Social Life in Medieval Rajasthan*, Lakshmi Narain Agarwal Educational Publishers, Agra, 1968, p. 227.

Raikas, a caste the duty to breed and graze them. According to this story, the origin of *Raikas* is traced from that day.'[111] Another saint, Tejaji, also sacrificed his life to protect livestock from invaders.[112]

THE ERA OF RELATIVE STABILITY

The history of Rajasthan from the eight century onwards till the emergence and consolidation of various regional principalities between the eleventh and fourteenth centuries is not very clear. One of the reasons for this tendency has been the relative instability of these Rajput principalities in their initial phase of state formation. As pointed out earlier, the environmental constraints had inhibited the growth of a large population base in these states. As a result, the economic base was not substantial enough to sustain these principalities. Not only were there continuous attempts to expand their dominion, at times they shifted to newer places in search of a more secure economic base. For example, Nainsi informs us that the Guhilots were initially based at Nagda, but later on, sensing the opportunity they settled in the more fertile and secure region of Chittor.[113] However, Shyamal Das places them in Valabhi.[114] Similarly, Rathors had also shifted their core area to more fertile and/or secure regions. Initially settled at Khed, near Pali, they flourished after occupying Mandore.[115] It was the same case with the Bhatis. They initially inhabited Lodurva[116] and later on, moved to the more strategic location of Jaisalmer.[117]

[111] Vinay Kumar Srivastava, 'The Rathore Rajput Heroes of Rajasthan: Some Reflections on John Smith's *The Epic of Pabuji*', *Modern Asian Studies*, vol. XXVIII, no. 3, 1994, pp. 597–9.

[112] Pemaram, *The Religious Movements*, p. 37.

[113] Nainsi, *Munhot Nainsi ri Khyat*, vol. I, p. 12.

[114] Das, *Vir Vinod*, vol. I, pp. 248–9.

[115] Ibid., pp. 12–25.

[116] Narain Singh Bhati (ed.), *Jaisalmer ri Khyat*, Rajasthani Shodh Sansthan, Jodhpur, 1981, pp. 46–7. Dasratha Sharma, 'Some Dates in the Bhattic Era and its Connected Problems', *Indian Historical Quarterly*, vol. XXXV, pp. 227–39.

[117] Nainsi, *Munhot Nainsi ri Khyat*, vol. II, pp. 6–8.

Moreover, since the ancient past, Rajasthan has never been a thickly populated region.[118] It is said that the region of Pokran was deserted in ancient times and was resettled by Ajaisi, the father of Ramdev Pir.[119] It is said that Mandore was sparsely populated when Rao Chunda was offered this territory to rule. He encouraged settlements in the region and developed it as the seat of an important regional kingdom.[120] It is evident from the pattern of settlement in the region as well. The population density has been low. Hence any process of reclamation of newer land demanded manpower to harness the geographical potential. To overcome this impasse', land was distributed among royal kinsmen so that they could colonize these lands and eventually, strengthen the resource base of the kingdom/region. Simultaneously, Rajputs started settling in marginal areas as well.[121] It can be argued that during the twelfth–thirteenth centuries the expansion and colonization of marginal areas was intensified due to political developments in north India. It has been pointed out earlier that the invasions from north-western India compelled the chieftains of the periphery to take shelter in the arid regions.

It has been aptly pointed out that, 'As these records suggest, at one level the process may have to be juxtaposed with a spate of colonization of new areas. The evidence of such colonization has to be traced not only in the significant expansion of the numbers of settlements but also in some epigraphic references, suggesting an expansion of agrarian economy.' However, it is important to recognize that, 'Any assertion about the increase in the number of settlements is, in the absence of any detailed historical–geographical study, only impressionistic. But in view of the widespread distribution of archaeological remains and epigraphs of the period as well as the appearance of numerous new place-names, there cannot be any doubt about the validity of the assumption.'[122] Such a large-scale colonization could not have been possible without

[118] Nainsi, *Marwar-ra-Pargana-ri-Vigat*, vol. I, p. 25; vol. II, pp. 2, 23 and 291.

[119] Ibid., vol. II, p. 291.

[120] Ibid., vol. I, p. 25.

[121] Ibid., vol. II, p. 1.

[122] Chattopadhyaya, 'Origins of the Rajputs', p. 62.

the growth of agrarian economy in the region[123] and in the arid region it was unimaginable without the intensification of irrigation potential. Initial impetus to this colonization may have come with the introduction of irrigation technology as discussed earlier.

B.D. Chattopadhyaya very elegantly points out other relevant dimensions of this very complex process of agrarian settlements in this relatively sparsely populated region when he suggests that, 'Rajasthan, in the period when Rajput polity was beginning to emerge, was in its various areas, undergoing a process of change from tribalism.'[124] This is evident from the references of conflicts between the aboriginal tribal elements such as Mers,[125] Minas, Bhils,[126] etc. Given the geographic features of the region, it is important to delineate this issue further as even till early modern times, it was difficult to sustain a pure agrarian economy in the region. Any disruption or deviation in the monsoon rains often led to reduced agrarian production and food scarcity. Till early modern times, larger polities, in order to retain the agrarian base and therefore, human resources, usually channellized resources to offer assistance for agrarian production.[127] Highlighting the significance of primary production units in the gradual emergence and consolidation of larger political edifice B.D. Chattopadhyaya has very eloquently examined in terms of supra-local or supra-regional polities.[128]

In the process of growing settlements in Rajasthan, the role of technological development needs to be reiterated. Newer developments in the technology and introduction of newer sources of

[123] 'If all this cumulatively suggests a proliferation of settlements, then the relationship of this process, through an expansion of agrarian economy, may be postulated with the emergence of the early Rajputs from about the seventh century', Ibid.

[124] Ibid., p. 64.

[125] Ibid., vol. I, p. 9.

[126] Ibid.

[127] Devra, *Rajasthan ki Prashashnik Vyavastha*; Gupta, *The Agrarian System*; and Singh, *The State, Landlords.*

[128] B.D. Chattopadhyaya, 'Political Processes and Structures of Polity in Early Medieval India,' in B.D. Chattopadhyaya, *The Making of Early Medieval India*, Oxford University Press, New Delhi, 1994, pp. 183–222.

draught power must have provided required impetus. S.P. Gupta has suggested that the introduction of 'dromedaries, or camels with single humps' and 'with the appearance of the pot-garland (*mala*) for the water, the mechanism was convenient more for lifting drinking water'; 'we are thus able to point to at least two significant technological developments that could have given a new economic profile to the Thar just on the eve of the middle ages'.[129]

The role of technology has been stressed in the penetration of the Rajputs into newer areas (inhabited or uninhabited). Gradual expansion of irrigation facilities and the resultant expansion and improvement of agriculture during this period suggest greater utilization of natural resources. Subsequent emergence of regional kingdoms and their expansion could not have taken place in the absence of agricultural surplus generated largely due to better irrigation facilities. A careful analysis of inscriptions of the period suggest that:

> it is perhaps significant that evidence relating to crop production and the emergence of settlements in water scarcity areas like Marwar does not date back to a period much earlier than the early medieval. This leaves room for postulating a connection between territorial expansion of agriculture and artificial irrigation. Secondly, the reference to double cropping, although it is the only one of its kind, would establish that a certain growth in production could be achieved through the organisation of artificial irrigation.[130]

The role of irrigation in transforming the nature of rural settlements is also evident from the later period also.[131] For example, village Singrampur in pargana Chatsu had been deserted for the past 100 years, however, with the construction of a non-masonry

[129] Gupta, 'Reconstructing the Political and Economic Profile', p. 161.

[130] Chattopadhyaya, 'Irrigation in Early Medieval Rajasthan', p. 51.

[131] B.L. Bhadani, 'Well-irrigation in Marwar in the 17th Century', *Shodh Patrika*, vol. XL, pt. 1, pp. 54–70; Das, *Vir Vinod*, vol. I, p. 118. Shyamal Das informs us that the practice of artificial irrigation was not very prevalent 200 years back. However, he wrote this in the very specific context of Mewar where rains and rivers sufficed for the purpose. Moreover, digging of wells in this hilly terrain was very cumbersome and capital intensive.

well agriculture flourished.[132] Thus it is evident that during this period a gradual expansion of agrarian economy was taking place.

In an area of surplus land, it is usually non-availability of either manpower or water for irrigation or both were responsible for lack of agriculture and settlement. To put it in different words it will not be inappropriate to suggest that paucity of investment; human as well as material, played an important role in relatively sparse settlements in the region. Thus as has been argued above we see a process of gradual peasantization of tribal[133] and gradual appropriation of pastoral community in seasonal activity of agriculture. Along with this one can cite other measures to accumulate necessary capital and material to fructify the settlements in the region. One of the important means to accumulate necessary capital was to loot the caravans passing through your territory and/or plunder the neighbouring thriving areas. Loot and plunder was in the form of gold, cattle and at times, even horses.[134] Rao Viram took possession of a caravan of horses which was coming from Gujarat.[135] According to tradition, a local goddess, Chanwana Ji, instructed Rao Chunda to raid a caravan which was carrying gold. He followed the instructions of the goddess and got the money. This wealth was quite helpful in his attempt at consolidation of his rule and later on, expansion.[136]

THE ERA OF EXPANSION

The process of territorial expansion had another very interesting feature. There seems to be a stress on colonization of the newer usually marginal land. Furthermore, it is interesting to note that measurement unit for the land in the land grants was in terms of

[132] Singh, *The State, Landlords*, pp. 51–2.

[133] Nandini Sinha Kapur, *State Formation in Rajasthan: Mewar during the Seventh-Fifteenth Centuries*, Manohar, Delhi, 2002.

[134] Nainsi, *Marwar-ra-Pargana-ri-Vigat*, vol. II, p. 119.

[135] Ibid., p. 19.

[136] Ibid., p. 2; Norman P. Ziegler, 'Evolution of the Rathor State of Marvar: Horses, Structural Change and Warfare', in Karine Schomer, Joan L. Erdman, Deryck O. Lodrick, and Lloyd L. Rudolph (eds.), *The Idea of Rajasthan: Explorations in Regional Identity*, vol. II, Manohar, New Delhi, 1994, pp. 192–216.

hal units. One *hal* covers around 50 *bighas* of land. Since land was available in abundance it was measured in *hals*-land which could be cultivated. As we have pointed out earlier, unclaimed land was available in plenty (*Khali Desh*);[137] the need of the hour was human intervention. We have several evidences of colonization or settlement of a piece of land, which was once habituated but later on deserted. Nainsi informs us that Phalodi was a deserted region, and could not be colonized even during the reign of Rao Chunda, Rinmall or Rao Jodha. It was during the reign of Rao Suja, that he ordered his son Nara to go and settle this *Khali Desh*.[138] Similarly, he informs us that though Merta was an ancient city, it was later deserted. Bushes had grown all around the old habitation. However, during the reign of Rao Jodha, it was once again colonized. He asked his sons, Var Singh and Dooda, to go and settle the region.[139] Similarly, we have evidence for Pokaran. Earlier, it was occupied but later it remained deserted for a very long period. Later on the father of Pir Ram Deo settled it.[140] At the moment it will be inappropriate to conclusively suggest that these areas did not witness settlements during intermediary period. The paucity of archaeological excavations and other historical evidences makes it very difficult for me to offer a cohesive account of situations during the intermediary period.

The process of clan-state formation in such harsh environmental conditions was not an easy process. The resource deficiencies were at times compensated as discussed earlier, by loot and plunder. In this process of expansion of the territory, the sons and brothers of the ruler were made to occupy the various territories conquered[141]/settled[142] by them. These territorial holdings were known as their *thikanas* and this practice was called *bhai-bant*.[143] Following *bhai-bant*, Rao Jodha assigned the conquered territories to his sons and

[137] Nainsi, *Marwar-ra-Pargana-ri-Vigat*, vol. II, p. 1.

[138] Ibid.

[139] Ibid., p. 37.

[140] Ibid., p. 291.

[141] Nainsi, *Marwar-ra-Pargana-ri-Vigat*, vol. I, p. 39

[142] Ibid., p. 25; vol. II, pp. 1, 37 and 291.

[143] Nainsi, *Munhot Nainsi ri Khyat*, vol. II, p. 120. Please see also, Sharma, *Rajput Polity*, pp. 4–5.

brothers who had fought either on his behalf or independently.[144] The territories of Khinvsar, Asop, Bikaner, Kherwa remained with his sons, Karamsi, Raipal, Bika and Samantsi, who conquered these areas and established their *thikanas* at these places respectively. The process of expansion continued with some setbacks in-between till the advent of the Mughals in north India. Later on, the Mughals played a crucial role of influencing not only the territorial distribution but also the nature of territory distribution among various clans/descendants.

Jodha's son Bika led an expedition into the region of *jangal,* which was then occupied by various tribes. The Bhatis and the Jats measured swords with him and were defeated. He extended the territory of the Rathor kingdom, but very soon became independent of the Rathors of Jodhpur. He extended his state to the southern limits of the Punjab. His advent marks the commencement of the rule of a new dynasty based in the present Bikaner region. The dynasty endured for over 500 years. Later on, they accepted the suzerainty of the Mughals, when he (Rao Kalyan Mal), together with his son (Rao Rai Singh), entered the emperors' service.[145] However, the political history of the period is not very clear.[146]

The political history of north-western Rajasthan becomes more legible with the advent of the Bhati Rajputs in the eleventh century. The Bhati king Devraj pledged himself to bring civilization and culture to this barren land. He also assumed the title Maharawal. Later on, Maharawal Jaisal shifted the capital from Lodurva.[147] The *Bhatti-vamsha-Prasasti* states that the valiant Jaisal found a new city in a place which was full of thorny shrubs and scrub, jungles, having *aka, thon, neem, jal, ker, baval, khejra* and many other trees.[148] It also states that the foundation of the town took place on a Trikut hill.[149] The new fort was erected in about AD 1156.

[144] Sharma, *Rajput Polity*, p. 5.

[145] Abul Fazl, *Ain-i-Akbari*, vol. I, p. 384.

[146] Ibid., pp. 474–5.

[147] Bhati, *Jaisalmer ri Khyat*, pp. 46–7.

[148] Nainsi, *Munhot Nainsi ri Khyat*, vol. II, pp. 6–8.

[149] Bhati, *Jaisalmer ri Khyat*, pp. 46–7.

The history of Jaisalmer, henceforth, has been one of in-fights or conflict with the neighbourly Muslim states of Sind and Multan and new emerging principalities of Jodhpur and later on Bikaner.

As far as central and the south-eastern Rajasthan are concerned, it was already densely inhabited, so in this region the power shuffled between the various clans of the Chauhans. Ananta-gochar is mentioned in the Harsha inscription as the land of the early Chauhans. It seems to be the area which had Nagapura or Nagore as one of the chief towns and which included Harsha as well as Sakambhari. Anant being the name of a snake king, the word Ananta-gochar probably meant nothing more than Nagaksetra, by which or a similar name this territory might have been known.[150] A closer look at the nomenclature presents a different picture of the landscape and settlement patterns of the region. Gochar literally means pastureland and thus it will not be inappropriate to suggest that this area must have had once sustained good amount of pastoral activities and communities. Furthermore, it also substantiates my argument that in this region agriculture was to a great extent supplemented by animal husbandry. Last but not the least one needs to bear in mind that agrarian settlements in the marginal areas were always fragile and predominance of either agriculture or pastoralism remained very fluid.

The history of settlements, emergence, expansion and consolidation of various kingdoms in different parts of Rajasthan is clear enunciation of continuous negotiations with variable waterscapes. Waterscape played a very important role in this process. The south-east, south, north-east areas being more wet necessitated relatively lesser investments in irrigation and resulted in early and more continuous settlement. Relatively better provisions for irrigation prompted early immigrants to settle in this region. This is the region which is more densely populated compared to other regions of Rajasthan even to this day. Furthermore, an economically viable political structure in the arid region could be sustained only with the development of agriculture in this region. The clan-based system could not emerge as a centralized political

[150] Sharma, *Rajasthan Through the Ages*, vol. I, p. 17.

structure due to very limited production possibilities. The surplus generation was very meagre, thus, we rarely find the emergence of a strong polity. At the same time the most significant element in the Rajput political system was the clan and clan loyalties superseded all other kinds of political affiliations. Nevertheless, the emergence of Rajputs and their later consolidation played a very important role in the expansion of settlements to the whole of Rajasthan.

THE ERA OF CONSOLIDATION

The situation changed only with the emergence of a centralized empire under the Mughals whose resource base lay in the more fertile regions of the Ganga–Yamuna *doab*. Economic development of the region was encouraged by the integration with the Mughal empire. A similar trend was visible in other parts of the globe as well. Arlene Miller Rosen argued that 'extension of farming zones into marginal areas in Near East' can be explained by its integration with the larger economies of the empires of the time. Extension of agriculture in marginal land could be sustained because of 'empire-wide trade and exchange' and 'its role in buffering the subsistence systems'.[151] A centralized empire by establishing comparatively greater peace and security, by enabling trade and commerce to expand, by increasing and diversifying the purchasing power of the consuming classes and thus helping the development of industries and manufactures, brought out conditions favourable to the growth of money economy. The emergence of money economy began to considerably affect agricultural production, especially because revenue was being collected more and more in cash.[152] It also led to the expansion of cash crops and the extension of the cultivated area partly as a result of the demand for greater revenue. The Mughal emperors also compelled the autonomous chiefs to conform to imperial regulations, especially in regard to the maintenance of law and order and the freedom of transit. The ruling chiefs found it

[151] Rosen, *Social Responses*, pp. 514–56.

[152] Habib, *The Agrarian System*.

more profitable and advantageous to join the imperial service than to attempt to retain their status as a big ruler.[153]

The political integration of Rajasthan into the Mughal empire had far-reaching implications for the Rajput clan states. The administrative reorganization of the territory of the Mughal empire, that is, creation of *subah*, *sarkar* and *pargana* boundaries gave an identity to Rajasthan as a region. The majority of the *parganas* falling in Rajasthan were included in the Mughal *subah* of Ajmer yet the various sub-regional identities were not altogether absent, however, sub-regional identities were never crystallized. It is therefore difficult to reduce Rajasthan to a singular regional identity. The subsequent histories of the various regions have been confined to their acceptance of Mughal suzerainty because henceforth, these states tried to emulate the administrative set-up and other prevalent practices of the Mughals.[154] The practice of territorial expansion as pursued by Rajput principalities was supplemented with greater stress on intensive cultivation by bringing more and more land under cultivation and the cultivation of superior crops.[155]

The above discussion make it amply clear that growth of settlements in the more arid regions without the support of a perennial source of water was made possible by exploiting the benefits of both the pastoral way of sustenance along with the agrarian potential of the region. This negotiation also reflects an astute understanding of the dual character of water. If periodicity of water demonstrated consistency, then agrarian production was relied upon but if inter-annual or intra-annual flux dominated, then, the pastoral component took care of the sustenance. Moreover, delayed or untimely rains might be a cause of concern

[153] Saiyid Nurul Hasan, *Thoughts on Agrarian Relations in Mughal India*, People's Publishing House, Delhi, 1973, pp. 34–6.

[154] Singh, *The State, Landlords*; Sharma, *Rajput Polity*.

[155] According to a *firman* issued by Akbar to the governors, they were asked to promote agriculture, and to conciliate the *raiyat* (peasant) and take measures to distribute *taqavi*, so that the number of hamlets, villages, towns and cities could be increased. Satish Chandra, 'Society, Culture and the State in the Medieval India', *Historiography, Religion & State in Medieval India*, Macmillan, Delhi, 1996, p. 222.

for the agrarian production but rain would always lead to growth of vegetation and could be exploited by the livestock and camel. Moreover, the pastoral component also enhanced the possibilities of movement in times of stress to the better-watered areas and return with the restoration of a more favourable waterscape. Migration to wetter regions was and is still relied upon.[156]

Equally important has been enunciation of process which facilitated human adaptations to the supposedly increased aridity. It can be seen in varied responses beginning from abandoning their settlements, mostly urban, and dispersal over a wider area. There are also evidences that human habitations grew in climatically more challenged areas and these cases have been at the centre of investigation. Introduction and/or development of new technologies offering better adaptation to the changes in the ecological contexts coupled with reliance on age-old and often forgotten traditions offered necessary resilience to these societies to grow and consolidate in a variable climate. Equally important have been the diverse sustenance strategies practised among these 'successful' habitations. It is now a well-recognized factor that 'particularly vulnerable were the more specialized complex societies that had approached or surpassed local carrying capacity thresholds'.[157] What emerges from our investigation can be viewed in

> terms of a number of localized cultural schemata, each of which was configured at any point in time by its original composition, by the unique history of the human group that carried it, by the acquisition and incorporation of new information, and by the reorganization of old and new information into a coherent whole. Information that carried through from the past resided in the social memory of the group. New information was derived from many sources both internal and external to the system.[158]

[156] Singh, 'The Role of Mahajans'; Kumar, 'Situating the Environment'; Kavoori, *Pastoralism in Expansion.*

[157] Jeffrey S. Dean, 'Complexity Theory and Socio-cultural Change in the American Southwest', in Roderick J. McIntosh, Joseph A. Tainter and Susan Keech McIntosh (eds.), *The Way the Wind Blows: Climate, History, and Human Action*, Columbia University Press, New York, p. 104.

[158] Ibid., p. 105.

Last but not the least, the vagaries of the monsoon-related uncertainty and apt adaptation based on flexibility carried out by humans suggest that society had learnt the fact that stability needs to be fluid and understood in terms of flexibility. If you live in a monsoon-dependent region then for survival one should understand drought and make friends with the aridity.

CHAPTER 4

Waterscapes
Rain-water Harvesting and Agro-Pastoral Landscape

Earth is not called the blue planet just for nothing as its waterscape which to a very great extent determines the emergence and continuance of life forms in general and human settlements in particular. It is equally important to recognize, as has been argued in previous chapters, that the waterscape has been in continuous flux since the remote past. The survival of humans during the changes in landscape in general and the waterscape in particular is apt demonstration of the human capabilities of adaptation/s. Adaptations to changes were clearly manifested in the methods of human appropriation of the immediate environment especially with reference to the conservation of water, or to be more precise, rainwater. As has been pointed out earlier, water was available primarily during the four months of the monsoon. Here it is important to emphasize that in the process of appropriation of naturally available water, various geographical features were integrated to secure maximum possible benefits without disturbing the regeneration capacity. It can be seen in the way natural cycles—monsoons, the monthly lunar cycle, annual solar cycle—were understood and put to constructive purposes in conjuncture with the nature of the soil, the local topography and other related factors. Semi-arid parts of Rajasthan could rely more on agricultural production, whereas in the arid parts, pastoral activities dominated. The agriculturally more-conducive areas have sustained a variety of crops in both the seasons, that is, *rabi* and *kharif*. However, the less conducive ones have survived on a single crop a year and extensively supplemented their livelihood through pastoral activities and trade. It will not be out of place to reiterate that many different political and social factors have influenced the growth, development and sustenance of social formations over the centuries. Similar to the growth of a newer set of negotiations to the changes or developments in the technological or social domain, even minor changes or aberrations in the basic facets of the physical environment—land, soil, rainfall and natural vegetation—often resulted in new sets of negotiations. Nevertheless, the broader structure remained the same. The production possibilities kept on redefining themselves, as any introduction of technology better suited to the variability of the climate gradually but ultimately influenced the productivity of land.

The first section of this chapter provides the description of the methods of potable water management as practised in the late medieval period. Methods through which the needs of the cattle wealth of the region were taken care of have also been included in the section. The next section carries an analysis of water management systems for irrigation evolved in response to the possibilities of agricultural production. Towards the end of the chapter we have examined the issues of conflict and contestations related to water.

WATER SECURITY: A PREREQUISITE TO HABITATION

The monsoon, especially in Rajasthan, is confined to four months only, with some rains in winter due to the western disturbances over the Arabian Peninsula. It was necessary to manage the water thus received during the monsoons for potable as well as irrigation purposes. In Rajasthan, the water table varied a lot and was mostly brackish in the western part for most of the times. As one moves away from the Aravallis in the broad belt across the desert, the supply of water in the wells 'is constantly failing or turning brackish'.[1] In terms of waterscape we can divide Rajasthan into two broad sections, one east of the Aravallis and another west of the Aravallis as these regions are distinct not only in terms of rainfall but also in terms of physiographic features.

Given the limited rainfall and paucity of a perennial source of water people practised various indigenous methods of water conservation. Most of the methods were practised at the household level whereas some required larger community participation or the state's financial support. The most important method of water harvesting was the wells. With the variation in the depth of the water table, the structures and functioning of the wells also changed. The regions with deep water tables necessitated deployment of different water lifting devices. In this respect, the nature of requirement, potable or irrigation, was also an important consideration. There are several terms for wells, like, *chanch, kosita, dhimra*[2] or *dhibra, kohar, kuwa, bera* and *baori* (step well).

Wells

Along with the depth of the water table in the western part of Rajasthan, the brackish nature of water has been an important environmental constraint in agricultural expansion.[3] Similarly, the character of the subsoil, especially in the desert and mountainous part of Rajasthan renders the construction of wells or tanks expensive and labour-intensive. The cost of construction of brick-lined wells must have been an important consideration.[4] For the Marwar region, Nainsi carefully distinguishes between earthen and brick-lined wells.[5] Wells also required maintenance, for example,

[1] *Imperial Gazetteer of India*, vol. XXI, p. 84.

[2] Nainsi, *Marwar ra Pargana ri Vigat*, vol. III, p. 132.

[3] Bhadani, *Peasants, Artisans and Entrepreneurs*, p. 46. For the Marwar region, Bhadani concludes that the sparse distribution of wells in Pokaran/Phalodi was in line with the scarcity of underground water and its availability at great depth. This was more so probably because the brackish water made only one out of twenty wells useful for irrigation (p. 51).

[4] Powlett, *Gazetteer of Ulwar*, p. 90.

[5] Nainsi, *Marwar ra Pargana ri Vigat*, vols. II and III.

in the village Dhadahanriyo of pargana Jodhpur, a well became non-functional as it was covered with sand.[6]

In the desert part of the region, all along the banks of seasonal rivers, both perennial and seasonal, wells were numerous and least expensive. These were known as *seja* or spring wells under the impression that they had reached the source of the spring. *Akara* was the name given to those wells which are avowedly percolation wells, and the supply of water was more precarious than in *seja* wells. The well was filled by a more or less rapid system of percolation, the deepest and most expensive wells often ran dry after being worked for a few hours, and the owner had to wait till the supply was renewed.[7] The extent of land irrigated by each well in a season averaged about five *bighas* or slightly more than 2.5 acres.

As far as distribution of wells in the region is concerned, the physiognomy played a crucial role. This has been provided by Nainsi and summarized in Table 4.1.[8]

TABLE 4.1: WELLS IN MARWAR (*c.* 1658)

Name of the Pargana	Area (in sq. mile)	Total no. of recorded wells	Wells (per sq. mile)
Merta	1,9111.03	6,947–7,782	3.64–4.07
Pokaran/Phalodi	4,222.89	538–638	0.13–0.15
Sanchor	1,382.12	276	0.20
Jalore	343.95	693	0.20
Jaitaran	63348	1,789–1,883	2.82–2.97
Siwana	1,251.51	500–544	0.40–0.43
Sojat	1,357.73	1,441–1,522	1.06–1.12
Jodhpur	9,880.69	4,031–4,065	0.40–0.41

The sparse distribution of wells in Pokaran/Phalodi was in line with the scarcity of underground water and its availability at

[6] Nainsi, *Marwar ra Pargana ri Vigat*, vol. I, p. 296.

[7] Ibid., vol. II, p. 96; Boileau, *Personal Narrative*, p. 105.

[8] Bhadani, *Peasants, Artisans and Entrepreneurs*, p. 46. Bhadani's calculations are based on the information available in *Marwar ra Pargana ri Vigat*. However, for ease of reference the tables provided by Bhadani have been used.

a greater depth.[9] This was more so probably because the brackish water made only one out of 20 wells useful for irrigation. Thus, it can be said that not only the depth at which water was available, but also the quality of water was an important environmental constraint in agricultural expansion.

The character of the subsoil renders the construction of wells or tanks expensive and labour-intensive. The *Vigat* carefully distinguishes between earthen and brick-lined wells. In pargana Merta, out of 6,947 to 7,782 wells, only 20 wells were brick lined.[10] In Jalore, out of 693, only 147 were *pakka kuwa* (brick-lined wells). Thus the *kachcha* wells heavily outnumbered the brick-lined ones. The much higher cost of construction of masonry wells was perhaps a factor. For example, in AD 1738, in *qasba* Sawai Jaipur, the cost of a masonry well was Rs. 300 while a *kachcha* well was sold for Rs. 84 in a village of pargana Phagi in AD 1764. In pargana Dausa there were 513 wells distributed over 18 villages, out of which 478 were *kuchcha* and 35 *pukka*, nine of the latter being equipped with *dhenkali*—a water-lifting device.[11]

As pointed out earlier, wells also varied a lot depending upon the depth at which water was available, the kind of water-lifting devices attached to it, source of water, uses, etc. Nainsi has described these as *chanch, kosita, dhimra*[12] or *dhibra, kohar, kuwa, bera*, etc. However, he has combined well and water-lifting devices. The *chanch* is a shallow well with the *dhenkli*[13] mechanism. The device of *dhenkli* was based on the principle of the lever, and could only be used on shallow wells where the water level was not more than 6 to 9 ft. below the surface. The *kosita*[14] well most probably had the *charas* operated on it. Presumably *kosita* was not a shallow well, since two different apparatuses, *charas*, and sometimes, the Persian wheel were used. Nainsi records the depth of water for *kosita* wells with a Persian wheel (45 to 84 ft, recorded in terms of

[9] Bhadani, *Peasants, Artisans and Entrepreneurs*, p. 51.
[10] Nainsi, *Marwar ra Pargana ri Vigat*, vol. II, pp. 89–213.
[11] Singh, *The State, Landlords and Peasants*, p. 52.
[12] Nainsi, *Marwar ra Pargana ri Vigat*, vol. III, p. 132.
[13] Bhadani, *Peasants, Artisans and Entrepreneurs*, p. 42.
[14] Nainsi, *Marwar ra Pargana ri Vigat*, vol. III, p. 130.

hath[15] and *puras*[16]), and without a Persian wheel from 15 to 45 ft. Thus, Nainsi's evidence indicates that where water was comparatively near the surface, *charas* was used and where the water level was much lower, recourse was take to the Persian wheel (Plate 1 A & B).

Dhimra/Dhibra also carries two meanings, that is, a well and Persian wheel. Nainsi specifically mentions the combined word as *arhat, dhima,* and sometimes splits it into two. It can be inferred that *dhima/dhibra* was a kind of well with a Persian wheel. The water level of *dhimra* wells given by Nainsi (42 to 84 ft or 7 to 14 *puras*) confirms that only this apparatus could have been used effectively for drawing water. The *kohar* and *kuwa* were also deep wells mostly used for drinking water, though sometimes sparsely used for irrigation as well.[17]

Apart from conventional wells, which were the most common method of water harvesting in north India, several other methods of rainwater harvesting were practised. One such method of water harvesting known as *kuin* or *par* or *teeni*[18] has been discussed in previous chapter. This system was practiced in a relatively small section of Rajasthan, which broadly corresponded to the area where sand dominated the landscape like, Churu, Bikaner, Jaisalmer and Barmer districts. This technique speaks for the better appropriation of given environmental conditions and better management of limited natural resources. The presence of an underground layer of limestone ensured that rainwater would remain potable and could be harnessed. Moreover, sand does not bind up like soil, hence, restricting the possibilities of evaporation of trapped percolated water.[19] Rainwater so retained by sand gradually seeps into the open space available at the bed of the *kuin*. It was usually not more than three to four earthen pots in a day. Therefore, a *kuin* was not

[15] Ibid.

[16] Nainsi, *Marwar ra Pargana ri Vigat*, vol. II, p. 14; *Munhot Nainsi ri Khyat*, vol. I, p. 349.

[17] Bhadani, *Peasants, Artisans and Entrepreneurs*, p. 42.

[18] Nainsi, *Marwar ra Pargana ri Vigat*, vol. II, p. 96.

[19] Misra, *Rajasthan ki Rajat Boonden*. He has provided the details of the method of construction of such wells.

A

B

PLATE 1: PERSIAN WHEEL STILL IN USE IN MEWAR REGION

meant for round-the-clock water availability[20] and was used for drinking purposes only. Nainsi informs us that village Bhojavas of pargana Jodhpur had a *teen*, which could supply water only twice in a single day.[21] The prevalence of the *kuin* highlights the resilience and adaptive capacity of the society.

Similarly, *bera* or *beri*, another method of water harvesting reflects the social adaptation to the environmental constraints of the region. It is a variation of a well, based on a similar principle as of a *kuin*. *Bera* were constructed near a water body or on the dry bed of a water body. The permeation from any water body to the adjoining area provided an opportunity to extract trapped water.[22] It is extensively practised even today in western Rajasthan. Nainsi has recorded that village Godhali of pargana Phalodi had a *beri* near seasonal stream.[23] Similarly, village Naberi of pargana Phalodi used the dry bed of a seasonal stream for agrarian production.[24] On the other hand, Nainsi informs that the groundwater in the village Bhabhclai of pargana Jodhpur was brackish. The village was without any source of potable water, but a *beri* constructed on a nearby tank sustained the habitation.[25] Similarly, village Mahev in pargana Sojat had a *bera* in the *talab* of the village and provided water when the *talab* dried up during the summers.[26] Village Chanda Vasani also had a *bera*.[27] In Jaisalmer the bed of the water tank, Amarsagar, sustained several *beri*. Amarsagar tends to retain water for only six to eight months and for the remaining part of the year these *beri* were the only source of water.[28] These techniques were prevalent in the regions with deep groundwater and limited possibilities of retention of water in the ponds for more than a few months. (see Plate 2). A similar principle was practised

[20] Nainsi, *Marwar ra Pargana ri Vigat*, vol. I, pp. 313 and 327; vol. II, p. 96.

[21] Ibid., vol. II, p. 313.

[22] Boileau, *Personal Narrative*, pp. 33–4, 37, 49 and 105.

[23] Nainsi, *Marwar ra Pargana ri Vigat*, vol. II, p. 21.

[24] Ibid., p. 25.

[25] Ibid., vol. I, p. 263.

[26] Ibid., p. 457.

[27] Ibid.

[28] Misra, *Rajasthan ki Rajat ki Booden*, pp. 56–7.

PLATE 2: BERI IN A POND

to manufacture salt in the regions with brackish groundwater. In pargana Pokaran, there was a *rin*, a salt land. The water procured through the *beras* was spread in fields to dry up[29] and after several stages salt was obtained.

Step Wells

Another form of wells quite prevalent in western India has been the *baori* or the step well. As is apparent from the nomenclature, *baori* that it is a well with steps. Usually, given the greater depth of the water table in the region, steps were provided to facilitate fetching the water. The basic principal behind the functioning of *baori* or the step well was the same as of the wells, that is, to tap the ground water. Though, argued in terms of nomenclature, step wells were rarely used for irrigation purposes. V.K. Jain[30] has argued that the word *vāpi* is derived from the Sanskrit word *vap*

[29] Nainsi, *Marwar ra Pargana ri Vigat*, vol. II, p. 36.

[30] V.K. Jain, *Trade and Traders in Western India*, Munshiram Manoharlal, Delhi, 1990, p. 27.

which means to sow,[31] thereby it is argued that *vāpi* were used for irrigation purposes. Unfortunately, on the basis of nomenclature, it is erroneous to argue that step wells were basically used for irrigation purpose. Jutta Jain has argued that although on the basis of inscriptions *vāpi* is translated as an irrigation well, it had various other socio-religious functions to perform.[32] On the basis of his extensive survey of the extant water bodies and inscriptions, C. Margbandhu is also of the opinion that step wells combined utilitarian and social functions.[33] However, unlike a tank, a step well does not provide adequate facilities for activities like bathing, cleaning and other ancillary activities.

Architecturally a step well has three major components—*kupa* (well), stepped corridor and *kuta* (pavilion). The major portions of the step wells lie beneath the ground surface. *Kupa* or a vertical shaft is the main component of a step well usually equipped with an arrangement for directly fetching up water by buckets (*Ghaya yantra*).[34] The next integral part of any step well is the stepped corridor. The stepped corridor leads down to several stories depending upon the water table in the region. A corridor with steps begins from the entrance pavilion and goes deep till the base where it merges with the rim of the well. In most of the cases these corridors are richly carved and ornamented with decorative relief and sculptures. At times these corridors include numerous intermediate towers like pavilions (*kutas*) built as open halls (*mandapa*).[35] Generally the entrance pavilion is a raised platform

[31] Monier Monier-Williams, *Sanskrit-English Dictionary: Etymologically and Philologically Arranged with Special Reference to Cognate Indo-European Languages*, Clarendon Press, Oxford, 1951, p. 941.

[32] Jutta Jain-Neubauer, *The Step Wells of Gujarat*, Abhinav Prakashan, Delhi, 1981, p. 11.

[33] C. Margbandhu, 'A Late Medieval Step-well at Nimarna, District-Alwar', in P. Chennareddy (ed.), *An Archaeological and Cultural Study: Researches in Archaeology, History and Culture in the New Millennium*, Sharada Publishing House, Delhi, p. 51.

[34] Ibid.

[35] Bharati Shelat, 'The Step-well Inscription of Gujarat: A Cultural Study', in P. Chennareddy (ed.), *'Krishnavinandan' Archaeological, Historical and Cultural Studies* (Festschriftt of V.V. Krisnasastry), Research Press, Delhi, p. 175.

which can be approached from three sides by the steps and the fourth side being the corridor. Thus, the structure allows one to descend down to the water level and during times of scarcity one can descend down to the shaft of the well located at the base of the corridor. The number of terraces and steps depend a lot on the depth of the water table of the location.[36] This kind of structure facilitates one to use water without much effort even if it is available at a very low level. Another significant and integral feature of the step well was the possibility of reaching to the water straight from the outer boundary of the well. This side usually also had a provision for a pulley to facilitate one to fetch the water from the great depths of the well.

The method of construction of its platform, galleries and ledges providing access to all the stories indicates that its interior provided such facilities. The steps of *baories* are constructed in different designs. The simplest and most common construction is one in which steps descend from the ground surface (level) to the water level which is available at the lower-most level, as one may find while entering in to a dungeon. The steps as well as levels (tiers) of each storey are so constructed that water could be accessed at every level. As one descends, one realises that the width of the steps starts tapering and the ambient temperature starts dropping.[37]

Other designs are generally with the steps in the shape of the English letter 'L' or where the steps leading to the platform form both sides like a 'T' and then the single flight of stairs turns at a right angle towards the water sources. Steps have been designed in an inverted 'V' of the English alphabet to allow a gradual descent from the three wall faces except the draw-well wall.[38] At times, to mitigate the higher inclination these structures had a unique scheme of arrangements for the steps. Each vertical descent was followed by a horizontal descent. As a result greater depth could be accessed without increasing the length of the corridor.

[36] Prabhubhai K. Patel, 'Construction and Structure of Step Wells', *Puratattva*, vol. XXVIII, 1997–8, p. 90.

[37] *Step Wells of Rajasthan*, Department of Tourism, Art, and Culture, Govt. of Rajasthan, Jaipur, 1998, p. 7.

[38] Ibid.

Structurally, step wells depend on an underground aquifer, often with a deep water table. Step wells are generally not feasible in regions with adequate rainfall; nor are they practical where the water table is too far below the ground. With the increase in depth of the wells, the possibility of the collapse of the side-walls multiplies. Most step wells in Gujarat range in depth from 60 to 80 ft. Near the desert in northern Rajasthan, the water table can lie as much as 300 ft below the surface, and people must rely on cisterns or simple cylindrical wells.[39] However, there are several factors which are necessary to be kept in mind before any discussion on the emergence and spread of step wells in western India. The primary ingredient used in the construction of step wells is stone. These are executed by expert masons. In a sense these are *pukkah* as pointed out by Morna Livingstone: '*Pukkah* is the adjective for something durable and able to be cleaned, and these permanent materials are reserved for only the high castes.'[40] It is very evident from the style, material used and expertise required for the construction that step wells are a very cost-intensive method of water management and could not have emerged and survived without extensive support from wealthy patrons. Working with stone especially in a circular pit intended to retain water was a highly skilled job and expected a certain level of expertise. Working with a water body necessitated a higher level of proficiency in terms of execution to counter the possibilities of seepage and thereby defeating the whole purpose.

As a backdrop, it is important to delineate that till the fifth century, temples in India were made of wood for obvious reasons. It was easy to procure wood and it was not very difficult to work with. However, they were prone to losses due to fire; an integral part of worship. Another important consideration, given the tropical climate of India, had been the relatively smaller life-span of constructions of wood. Simultaneous developments in the ever-growing expertise in stone sculpture especially with the growth of the Mathura School of Art must have equipped the mason to work

[39] Morna Livingstone, *Steps to Water: The Ancient Stepwells of India*, Princeton Architecture Press, New York, 2002, p. 10.

[40] Ibid., p. 9.

with stone. Shashi Bhushan Singh has argued that by the middle of the fifth century AD, the art of sculpture in stone had grown local roots.[41]

> Although sedimentary stone eventually hardens, when it is fresh from the quarry it is almost as easy to carve as wood. The blocks of sandstone and tufalike travertine, introduced to build longer-lasting temples, proved ideal for water buildings. . . . [Moreover,] 'The Guptas' reach created a vital culture over the whole of north India, in part because it introduced "the full dignity of squared ashlar"—the technique of fitting stones together so tightly that they make a perfect fit in the wall with no cushion of mortar. The innovation of this method of stone construction made stepwells possible.[42]

It is interesting to trace the roots of India's expertise with stone. Morna Livingstone further argues that:

> Stones of unmortared blocks as a post-and-beam system was previously used on the limestone plateau of northern Syria, the main centre of stone building in the ancient Western World. Stone was used in Syria to make Roman like roads, aquaducts, mills, baths, and basilicas. Stepwells were not part of that tradition, but many construction similarities suggest a link between Syrian and Indian masons at the time of Dhank stepwells and perhaps two hundred years thereafter. (More than 800 Syrian Christian settlements—now called the *civitates desertae*—were left empty after the Muslim conquest of AD 636 when local workmen migrating in search of new patrons left behind "a stony silence.") This is especially evident in the way columns and lintels were used; the Syrian practice of extending gigantic stone blocks to form beams and thick stone ceilings became the standard for Hindu builders. Although stone technology may have come to Gujarat from temple workshops in central India run by the Guptas, construction technology and architectural ideas could have come to India separately.[43]

The epigraphic evidence of step wells can be located from the early period. As argued earlier, step wells were known as *vapi* and the references can be located in inscriptions, copper-plate grants

[41] Shashi Bhushan Singh, *Sarnath Ki Baudh Murtikala*, Swati Publications, Delhi, 2009.

[42] Livingstone, *Steps to Water*, pp. 9–25.

[43] Ibid., p. 25.

and also in *Śilpaśāstra*, a Sanskrit text on architecture. Epigraphic reference of *vapis* is sparse. The earliest epigraphic reference of step wells may be traced to Aśokan inscriptions offering a blurred reference of step wells. The seventh pillar edict of Aśoka mentions that wells provided with a flight of steps for descending into water were built by him.[44] The *Gangadhar* inscription,[45] found at Jhalawar 480 VS (CE 423) states that Mayuraksha, a minister of Vishwa Varma, got a Vishnu temple and a step well constructed; this can be treated as one of the earliest epigraphic evidences of step wells from the region. The following are the other inscriptions associated with the early phase of step wells of Rajasthan: A rock edict dated approximately VS 741 (CE 684), found at Nagar in Jaipur district, mentions that a step well was constructed here by the architects and sculptors of Marwar–Bhinmal region, suggesting that the Marwar–Bhinmal region had better masons more capable in the art of construction of step wells.[46] For the subsequent year also we find an inscription on the rocky surface in Kutila script (CE 685) at Mandore (Jodhpur) which mentions the construction of a step well.[47] These early evidences suggest that step well construction gained momentum in the late seventh century and it corroborates our hypothesis that the early medieval times witnessed a different set of human-nature interactions specifically with reference to water management both for potable and irrigation purposes. Soon the construction of step wells became an act of charity and religious devotion.[48]

[44] It is interesting to note that as per the 8th Pillar Edict, Aśoka suggests that at an interval of 8 *kos* wells were dug by his orders and flights of steps (for descending into the water) were built. E. Hultzsch (ed.), *Corpus Inscriptionum Indicarum*, vol. I, *Inscriptions of Asoka*, Archaeological Survey of India, Delhi, rpt. 1991, p. 130.

[45] G.H. Ojha, *Rajputane ka Pracheen Itihas*, Rajasthani Granthagar, Jodhpur (1st pub. 1926), 3rd rev. edn., 2009, p. 226; R.C. Agrawal, 'Rajasthan Ke Katipaya Prachin Sarovar Tatha Vapi', *Varda*, vol. VI, no. 4, 1963, p. 57.

[46] Agrawal, 'Rajasthan Ke Katipaya Prachin Sarovar Tatha Vapi', p. 53; Margbandhu, 'A Late Medieval Step-well at Nimarna, District-Alwar', p. 53.

[47] Gopi Nath Sharma, *Rajasthan ke Itihas ke Srota*, pt. 1, Rajasthan Hindi Grantha Akadami, Jaipur, 1983, p. 50.

[48] A step well called Lahana in *Basantagarh* (Sirohi), carries an inscription

The earliest structural evidence of a step well in India, so far known to us, is found at Dholavira, a Harappan settlement in the Kutch region of Gujarat. A rock-cut well with steps has been excavated in the central part of the eastern reservoir. The well was dug into the rock-bed. There is a shaft of chiseled steps to go down into the well to fetch water. The bed rock might have been excavated by the proto-historic people here up to the water table for ensuring regular water supply throughout the year.[49]

Jhilani (between AD 550–75) and Manjushri (about AD 650) wells in the village of Dhank in Saurashtra are termed as the first true step wells of the early medieval times. Dhank's wells are different from the earlier Harappan step wells and reservoirs, the Buddhist baths, and variously dated rock-cut water buildings in very fundamental terms. They are built of stone blocks laid without mortar and contain built-in shrines. Finding stone immune to the water must have been a great innovation at the time, for only this could have allowed builders to combine in one material stairs, shrines and a pool for drinking and bathing.

Given the present level of research in the field and the widespread distribution of step wells across the region, it is very difficult to map the process of evolution and spread of step wells in Rajasthan. On the basis of fieldwork carried out by the author and available literature, an attempt has been made to broadly outline the emergence and growth of the tradition of step well construction in Rajasthan. A very rudimentary form of step well with a very basic provision of steps can be located very close to the eighth century temple of Lord Ganesh in Kakuni, Baran district (Plate 3). It is difficult to suggest that the steps of this well could have been used

---

dated CE 1042, suggesting that a queen named as Lahani got this constructed as a work of charity and it was obviously named after her. Sharma, *Rajasthan ke Itihas ke Srota*, pp. 71–2; Another inscription of *Jhalarapatan* (Jhalawar) of the year VS 1143 (CE 1086) mentions that a person named Janaka who happened to be a *patel* during the reign of king Udayditya got this step well constructed. *Journal of Royal Asiatic Society of Bengal*, Calcutta, New Series, pt. 10, no. 6, 1914, pp. 241–3.

[49] R.S. Bisht, 'Dholavira: New Horizons of the Indus Civilisation', *Puratattva*, vol. XX, 1989–90, pp. 71–82.

PLATE 3: RUDIMENTARY FORM OF STEP WELL, JHALAWAR

to fetch water effectively. On the contrary, it can be suggested that these steps or projections in the well were never meant to be steps. These projections were created to facilitate the construction of a well in the hilly terrain of the region. The water table must not have been very low as a seasonal river, Parwan, flowed adjacent to it. Moreover, given the rocky terrain of the region digging a hole must have been a strenuous job and would have demanded a certain level of expertise. In general, Ravan ki Chanvari located close to Mandore railway station near Jodhpur is largely accepted as the earliest example of step wells in Rajasthan. It carries inscriptional details suggesting that the step well was constructed by Madhu, the son of Chanakya in the year VS 685 (CE 628).[50]

The tradition of step well construction reached a certain level of refinement by the time we come to the step well at Osian (Jodhpur) which belongs to eighth–ninth century CE.[51] It is also known as Katan Vav. It can also be described as *kunda*, combining the properties of a well with those of a reservoir.[52] It is rectangular in shape. There seem to be fifty steps to reach up to the water level. Unfortunately, today it is almost dry and in a dilapidated condition. Katan Vav, the great stepped pond at Osian, is one of the earliest examples of an early addition to a Jain and Hindu temple site. The pond looks magnificent in its orange-red stone. The pool has three walls of stairs, but the fourth wall houses shrines. There is extensive provision for pavilions.

> The pavilions in particular are beautifully carved. Osian's pool is a perfect study of the durability of the stepped pond's simple stair structure over the stepwell's more open roofs, columns, terraces, and pavilions resting on vertical supports. Here, three walls of stairs remain in usable condition, while the fourth wall of columns and pavilions is crumbling into ruin.[53]

Another similar but well-preserved step well is located at Abhaneri in Dausa district and is known as Chand Baori meaning 'moon' or 'silver' well. The step well is adjacent to the Hasratmata Mandir (temple). It is perhaps the steepest stepped pond ever built. Over the centuries, this step well has seen extensive additions apart from the regular ones like, occasional whitewash, mending the parapet, or reversing worn steps to ensure a flat tread, etc.

> . . . Abhaneri was refurbished over the centuries: the lower pool, shrines, open rooms near the water and stairs date from eighth century, but the upper stories were rebuilt as a palace around the eighteenth century by the Mughals, who also added a four-sided arcade with pointed arches,

[50] Agrawal, 'Rajasthan Ke Katipaya Prachin Sarovar Tatha Vapi', p. 58.

[51] Ibid., pp. 59–60.

[52] Julia Hegewald, 'Water Architecture in Rajasthan', in Giles Tillotson (ed.), *Stones in the Sand: The Architecture of Rajasthan*, Marg Publication, Mumbai, 2001, p. 81.

[53] Livingstone, *Steps to Water*, p. 38.

keeping the pool at the geometric centre. Thus Abhaneri shows two classical periods of water building in a single setting[54] (see Plate 4).

This step well has been dated to eighth–ninth century CE.[55]

PLATE 4: CHAND BAORI

The plan of Chand Baori demonstrates the spatial compactness of stairs triangles that run parallel to the pond's rim. The rise-to-run for each flight adds no more than eighteen inches of width to eight feet of drop, making an extremely sharp descent. The wall is so steep that the top of the stairs triangles, seen from the pond's brim, alternatively hide and reveal the people going down from above. . . . Its stairs are designed in the form of English alphabet "V" which are in fact constructed in the unit of several 4 to 5 stairs each. These stairs which give a look of "V" shape, as said earlier are present on the three sides of this water structure and on the whole is the special beauty of this baori or step well. Early stepped ponds such as the Chand Baori laid out in an eight-by-eight or nine-by-nine square grid did not refine their stairs, but instead used sharp, right angled blocks. In

[54] Ibid.

[55] Agrawal, 'Rajasthan Ke Katipaya Prachin Sarovar Tatha Vapi', pp. 59–60.

spite of this regularity, the Chand Baori uses seventeen different solutions to resolve the stairs-to-stair intersections at its corners.[56]

It is popularly believed that there is a tunnel in the reaches of the step well which leads to a *kund* known as Kulania Kund. Multi-storied lower verandahs of this step well are an example of structural beauty created by the Gurjar rulers. This step well was also provided with a pulley for fetching water. The device was supported on stone pillars.[57]

Nadol, a place in the district of Pali has fortunately offered us a few excellent specimens of technology and aesthetics associated with the construction of step wells, that is, Chuna ki Bowlee, Moong Vav, etc. However, Rupa Vav is considered as the most remarkable one among several others. It is believed that Rupa Vav was constructed during the closing years of the tenth century AD. The step well of Rupa Vav is structurally five storeyed and contains three pillared pavilions besides 26 steps for descending to approach the water (level), starting from the ground level.[58] Interestingly, unlike the dominant tradition, the side walls are devoid of any resting place. Its pillars are made up of plain flat stone slabs which support the roof following the principals of trabeate form of architecture; thereby, reducing the open space and creating small resting spaces. The space around the well is architecturally designed for standing a person or persons and it is covered by a stone canopy all around. The parapet of this well which is obviously on the ground level is provided with the arrangement of stone slabs and a support for the pulley for fetching water.[59] There is a debate over the feasibility of this well with a pulley for irrigation purposes. On the basis of terminology (as argued earlier) *vapi* signifies irrigational use, however, the structural location of the pulley contradicts this argument convincingly.[60]

[56] Livingstone, *Steps to* Water, p. 38.

[57] *Step-wells of Rajasthan*, pp. 15–16.

[58] Margbandhu, 'A Late Medieval Step-well at Nimarna, District-Alwar', pp. 53–4.

[59] *Step wells of Rajasthan*, pp. 19–22.

[60] Livingstone, *Steps to Water*, p. 38.

It is difficult to examine the architecture of these step wells purely in terms of functional utility. There is a debate on the nature of architecture visible in the conception of these magnificent water management systems. As far as the utility of the step wells is concerned, these structures supplied water to men and animals. Besides, these served as pleasant and refreshing retreats for the travellers as well as for villagers, especially during the summer months. We can broadly identify multiple layers of social relationships associated with step wells on the basis of their nomenclature. For example, *Śrāvak vāpi* means a step well financed by a follower of the Jaina religion; *Malla vāpi* can be deciphered as a step well constructed or occupied by a wrestler. Step wells were also identified by their location; a step well located near a *pipal* tree came to be known as *Pippal vāpi*. Similarly, *Shireesh vāpi* is a step well located near a *shireesh* tree—a kind of flowering tree or *Albizzia lebbek.*[61]

The tradition of step well continued[62] and later in the sixteenth and seventeenth century we find a growing practice of construction of step wells as a charitable endowment or in the memory of one's parent. Even the Rajput queens have been attributed with patronage for the construction of numerous step wells from their privy purse. The classic examples of this act can be seen in the construction of Rani Vav in Bundi and Hadi Rani ki Baori at Toda Raisingh. Toda Raisingh was the centre of a small principality ruled by the Solankis. It was during the late medieval times that the Hadas of Bundi were able to regain their control over Bundi with the help of the Solankis. Subsequently, the alliance was further strengthened by a matrimonial alliance when a Hada princess was married to

[61] Rajesh Jamindar, 'Agricultural Economy of Gujarat: Aspects and Appraisal up to 1300 AD, in L. Gopal and V.C. Srivastava (eds.), *History of Science, Philosophy and Culture in Indian Civilisation*, vol. V, pt. 1, *History of Agriculture in India (up to c. 1200 AD)*, Concept Publishing, Delhi, 2008, p. 739.

[62] The Sultanate period also saw a series of step wells. Step wells are located within or in the vicinity of shrines, for example: Dargah of Muin-ud-din Chisti at Ajmer, Nizam-ud-din Aulia, Delhi, Qutub area, Agrasen Baori, and numerous other. Anthony Welch, 'Hydraulic Architecture in Medieval India: The Tughluqs', *Environmental Design*, vol. II, 1985, pp. 74-81.

the Solankis. As a memorial and an act of charity a huge step well was constructed at Toda Raisingh and came to be known as Rani Hadi ki Baori. It was built in the seventeenth century. In design it is very similar to Chand Baori of Abhaneri. The prime difference between these two step wells being that on the water-drawing wall side in the Chand Baori there are a number of stories which might have been used for resting or as platforms from where the king and his entourage performed religious rites whereas, in the Hadi Rani ki Baori, there are large stories with resting halls which are extremely spacious and cool. This also offers, from the highest storey, an extremely picturesque view of the hills in the background (see Plate 5).

Rajasthan in a very peculiar manner offers the extant remains of the process of the evolution of step wells in the region from very rudimentary ones as described earlier and the highly developed ones like Rani Vav of Bundi. The adjoining region of Gujarat also offers extant specimens of step wells from very basic ones to highly ornamented ones as can be seen in Rani Vav of Patan. As argued earlier the beginnings can be traced to the close connections with the temple construction activity during and after the Gupta era.

PLATE 5: HADI RANI KI BAORI

However, the later upsurge offers a beautiful amalgamation of technologies usually associated with the Turko–Persian traditions as reflected in the style and materials deployed in the construction of the step well at Bandharej, etc.

Jhalaras

Jhalaras are another important method to harness the water potential of any given region. As argued, rainwater emerged as a major source of water to sustain the human settlement in the region especially during the early medieval times when several river streams either dried up or changed their course.

> Jhalars neither have any catchment of their own nor are they connected to any watercourse. They collect the subterranean seepage of a talab or lake located upstream. . . . Jhalars are essentially meant for community bathing and religious rites. Often rectangular in design, *jhalaras* have steps on three or four sides. The Mahamandir Jhalara, which is more like a *baori* is the oldest one, (in Jodhpur region) having been constructed in 1600.[63]

Kunds

The other important method of water conservation was *kund* or *kundi*. It was a mechanism based on the direct accumulation of rainwater even for drinking purposes. This technology of collection and retention of water was extensively used in north-western Rajasthan where groundwater was either brackish or available at great depths or both.[64] *Kundis* were constructed both for domestic and public purposes. The roof of the house was used as a catchment area to collect rainwater. Before the onset of rains the roof was cleaned and washed with the first showers. Subsequent rains were directed to the underground water tank. The underground tanks functioned as reservoirs and provided water in times of

[63] Anil Agarwal and Sunita Narain (eds.), *Dying Wisdom: Rise, Fall and Potential of India's Traditional Water Harvesting Systems*, Centre for Science and Environment, Delhi, 1997, pp. 124–5.

[64] Boileau, *Personal Narrative*, p. 13.

scarcity.[65] Even now, there are innumerable *kundis* in houses all over Rajasthan. Locally available lime mortar was used for the construction of *tanka* because it was able to restrict seepage.[66] Easy mechanism and minimal maintenance made it the most viable source of water for the lower strata of the society. All along the western region, one can still find numerous *kundis* existing in open fields for public use.[67] The open-field *kundi* also used a similar method to harvest rainwater. The roof of the underground tank itself functioned as a catchment area. An opening was made to enable one to move into the tank as well as to draw the water. At times, even the state promoted construction of such techniques.[68]

The bigger form of *kundi* is known as *tanka*. 'Occasionally where the wells were brackish, drinking water was supplied from *kundis* or pits in which rainwater was collected and preserved, but in such small quantities as to be used only in an emergency.'[69] The most important example of this mechanism is Karorpati Tanka of Jaigarh fort near Amber constructed in the fifteenth century.

> To the North-east of the city of Jaipur and beyond the Amber palace, on the spurs of the Aravalli range, is situated the fort of Jaigarh constructed by the Kachhawaha rulers of Amber. . . . Sawai Jai Singh (AD 1698–1740), before shifting his capital to his newly planned city (AD 1727), carried out major repair and remodelling of this fort. Strategically located the Jaigarh fort could monitor the defensive and offensive requirement of the ruler, . . . the most peculiar feature of this fort appears to be its hydraulic system which would not only have added to its basic strength, but also took care of its daily needs. Interestingly, striking similarities appear between the hydraulic systems of the ancient Roman and Palestine forts situated in the *wadi* and that of the Jaigarh Fort, that is, one finds water-channels and ducts, built all along the mountainous terrain outside the precincts of the fortification, conducting water during the spate of rains into the reservoirs and tanks constructed inside the fort.[70]

[65] Misra, *Rajasthan ki Rajat ki Booden*, pp. 56–7.

[66] Chand, *Tawarikh Jaisalmer*, p. 203.

[67] The religious merit attached with the construction of water storage systems for the public seems to be one reason for the greater proliferation.

[68] Chand, *Tawarikh Jaisalmer*, p. 200.

[69] Boileau, *Personal Narrative*, pp. 13–14.

[70] Rajiv Sharma and Syed Ali Nadeem Rezavi, 'Aspects of Hydraulic

Rajiv Sharma and Syed Ali Nadeem Rezavi have explained the functioning and technology of the tank very lucidly. The networks of the water channel commence on the hillocks and from the south; they form three loops before entering into the fort walls. These loops around the hillocks are formed in such a manner that rainwater falling on the hills is not allowed to drain down, but is collected in these channels to be directed towards the fort due to gravity flow (see Plate 6).

PLATE 6: KARORPATI TANKA

The covered subterranean channel veers its way through the second gate of the fort, and from these onwards, it is again in the open. This channel leads the water into a rectangular cistern having two sliced openings. Through the one, the water can directly enter into a large water reservoir, popularly known as *tanka*, which is wholly covered from above. Built in a square shape it has a capacity to hold almost 60 million gallons of water. The ceiling of this *tanka*

Engineering in Medieval Rajasthan: A Case Study of Water System in Jaigarh Fort', in Ahsan Jan Qaisar and Som Prakash Verma (eds.), *Art and Culture*, Publications Scheme, Jaipur, 1993, pp. 129–31.

rests on eighty-one arched pillars. On top of the ceiling and exactly in its middle is constructed a square platform with one opening. Towards the north, a couple of staircases descend below where a walkway/ramp is provided to draw water. These staircases can be appropriated through doors constructed towards the east and west.

If the water is not required to be stored into this large reservoir, the channel, through the second sluice of the cistern, leads it to a rectangular open tank. This tank, popularly known as the bathing tank, is also provided with a flight of steps leading down below.

Adjoining this open tank towards the south is the third tank, which, like the *tanka*, is covered with a roof. It is provided with nine holes on its ceiling to facilitate lifting of water from it. The surplus water from the open tank was carried into this covered tank with the help of circular sluices provided high above on the adjoining wall. This meant that only clear water would be carried into the third tank.

It was through this ingenious technique of capturing the rain-water on the hills and then directing it to the fort, that the Jaigarh fort was equipped to withstand a long siege. This technique also solved the problem of carrying water from the lakes situated far below the hillocks. Its efficiency can be gauged from the fact that even in present times, during the rainy season, water is carried through the same dilapidated and uncared for channels and viaducts. The water is still drawn from the *tanka* to cater to the needs of the tourists and the inhabitants of the fort.[71]

Reservoirs: Talabs, *Lakes and Tanks*

Creating a water body by constructing an embankment in the path of a river/rivulet has a long tradition in Rajasthan. We have a long list of water bodies/lakes constructed by individuals, communities and the state. Kings and nobles constructed numerous water

[71] 'The most peculiar feature of this fort appears to be its hydraulic system which would not only have added to its basic strength, but also took care of its daily needs.' Sharma and Rezavi, 'Aspects of Hydraulic Engineering in Medieval Rajasthan', pp. 129–31.

storage systems for drinking purposes.[72] A few prominent water bodies include the Udaisagar[73] built in AD 1564, Raj Samand constructed between AD 1662 and 1675 in Mewar[74] and Manasagar erected near Amber in the early eighteenth century.[75] The earliest man-made lake to be built in the Mewar region is attributed to a Banjara at the end of the fourteenth century. It is located in the capital, Udaipur. Subsequently, an embankment was raised by Rana Udai Singh in 1560. The lake is about 2.5 miles long by 1.25 miles broad, has an area of over one square mile and a capacity of 418 million cubic feet of water. In the middle stand the two island palaces, the Jagmandir and the Jagniwas, the former built by Rana Jagat Singh I (1628–52) and the latter by Jagat Singh II (1734–51).[76]

Udaisagar lies 8 miles east of Udaipur, being 2.5 miles long by 1.5 miles broad; its area is about 2 sq. miles, and it drains 185 sq. miles of country. The water is held up by a lofty dam of massive stone blocks, thrown across a narrow outlet between two hills, a little to the south of Debari at the eastern entrance to the Girwa or Udaipur Valley. The embankment has an average breadth of 180 ft and was built by Rana Udai Singh between 1559 and 1565.[77]

Maharana Raj Singh, the Sisodia Rajput ruler of the former princely state of Mewar (AD 1662–80), is credited with the construction of the Raj Samand as engraved on the eulogy posted on the embankment.[78]

The Raj Samand is situated about 36 or 37 miles north by north-east of Udaipur and just to the north of Kankroli. It is three miles long by 1½ broad, receives the drainage of 195 square miles and has an area nearly

[72] Nainsi, *Marwar ra Pargana ri Vigat*, vol. I, pp. 579–602.

[73] Nainsi, *Munhot Nainsi ri Khyat*, vol. 1, p. 30.

[74] Das, *Vir Vinod*, vol. II, pp. 578–634 and 2204.

[75] I.A. Khan, and Ravindra Kumar, 'The Mansagar Dam of Amber', in Anirudha Ray and S.K. Bagchi (eds.), *Ancient and Medieval Technologies in India*, Sundeep Prakashan, Delhi, 1986, pp. 25–40.

[76] K.D. Erskine, *Rajputana Gazetteers*, vol. 2-A, *The Mewar Residency*, Scottish Mission Industries, Ajmer, 1908, p. 109.

[77] Ibid., p. 9.

[78] Neelima Vashishtha, 'A Note on the Sculptures of the Raj Samudra Lake in Rajasthan', in Ahsan Jan Qaisar and Som Prakash Verma (eds.), *Art and Culture*, Publications Scheme, Jaipur, 1993, p. 111.

> three square miles. The lake is formed by a dam built at the south-western end by Rana Raj Singh-I between 1662 and 1676. Its construction served to alleviate the sufferings of a starving population and it is the oldest known famine relief work in Rajputana. It is said to have cost from 96 to 115 lakhs of rupees, or between £640,000 and £760,000. The dam forms an irregular segment of a circle nearly three miles long; the northern portion which lies between two hills, is about 200 yards long and 70 yards broad, and is entirely faced with white marble from the adjacent quarries of Rajnagar. Along the front, a flight of steps descends to the water's edge, while jutting out into the lake are three marble pavilions—two of sixteen columns each and one of twelve—all richly sculptured in different patterns.[79]

The reservoir does not have provisions for the canals, therefore the irrigation potential of this lake was very confined.

The tradition of development of big reservoirs by constructing a dam across the river/rivulet continued in Mewar and we see the construction of Jai Samand soon after. It is also known as Dhebar Lake. It

> lies . . . about thirty miles south-east of Udaipur and 969 feet above the level of the sea. Its length varies from one to five miles. It receives the drainage of 690 square miles and has an area of twenty-one square miles. On the west the hills rise from 800 to 1,000 feet above the level of the water, while the small wooded islands and the picturesque fishing hamlets on the northern shore add greatly to the beauty of what is one of the largest artificial sheet of water in the world. Jai Samand (*Jaya Samundra*—the sea of victory). The dam is 1252 feet long and 116 feet in height; its breadth at the base is seventy feet and at the top sixteen feet. . . . Behind the dam, at a distance of about hundred yards, is a second wall 929 feet in height with a breadth of thirty-five feet at the base and twelve at the top. The space between these two walls is being gradually filled in with earth.[80]

The tradition of building water bodies can be seen in other parts of Rajasthan. The only pre-condition being the topography which means a hilly terrain leading to the formation of narrow gorges. The narrow gorges between two hills on either side offered an ideal

[79] Erskine, *Rajputana Gazetteers*, vol. II A, p. 9.

[80] Ibid., pp. 8–9.

location for the construction of an earthen dam ultimately leading to the creation of a large water body/lake.

We have evidence of the construction of dams for Amber region also. There exist a number of dams, having roughly similar designs and dating back to the eighteenth century or earlier. One of these is the Mansagar Dam. It is located nearly 2 km north-east of the city of Jaipur. The tract at present occupied by the lake Mansagar created by this dam was formerly a marshy land. It also formed the bed of a rivulet which, passing through a narrow gorge in the hills, used to flow towards the north-east. This rivulet called Darbhavati was dammed in VS 1792/AD 1735. It may be further conjectured that the specialized knowledge relating to the problems of dam building that is so clearly reflected in the basic design of the Mansagar Dam was probably an indigenous tradition.[81] The study of Mansagar Dam suggests that it is an interesting specimen of a gravity dam built on the trabeate principle. The exclusive use of the trabeate principle in this dam tends to suggest that it represented a pre-Turkish tradition of dam building. This tradition appears to have survived in the territories dominated by the Rajput chieftains down to the beginning of the eighteenth century. Some of the structural features of the Mansagar Dam, like the exposed part of the plinth extending for several metres towards the lake, five bastion-like projections rising from the plinth, are possibly meant for countering the toppling or sliding of the superstructure as also for preventing a tension failure. This goes to indicate an awareness of the basic problems of dam engineering on the part of the architect who designed this dam.[82]

Lakes and dams played an important role as a source of irrigation in the Mewat region also. Babur describes extensively the Kotla Lake and noticed its utility for irrigation.[83] The other lakes of the region were Siliserh and the Deoti. As these lakes could retain water for only a limited period, their fertile dry beds were used for cultivation of crops.

[81] Khan and Kumar, 'The Mansagar Dam of Amber', pp. 25–40.

[82] Misra, *Rajasthan ki Rajat ki Booden*, pp. 19 and 54.

[83] Babur, *Babur Nama*, p. 580.

None the less, of singular significance were the construction of Garsisar and Jaitsar[84] in Jaisalmer, which exemplify a thorough knowledge of the topography of the locality. Garsisar is located on the foothills of Jaisalmer fort (see Plate 7 A & B). The location of Jaisalmer is significant in the sense that the region was stony and situated on the side of hills. A water body could be constructed by erecting a *bund* to block the drainage of the catchment area. Thus, if by any mechanism, runoff water could be stored, then the possibility of a big water body was present. This scheme was realized with the construction of Garsisar at the foot of Jaisalmer.[85] The construction of Garsisar could tackle the requirement of water for a capital city; it could also address the predicament of permanent capital.[86] Furthermore, Anupam Misra has explained that overflow from Garsisar was taken to seven subsequent reservoirs which ensured its utmost utilization. It also ensured the continuous survival of Jaisalmer in the arid region of Rajasthan.[87] One of the subsequent water bodies was Amarsagar. It was built in the fifteenth century AD. As the catchment of this water body was not very large, it could not retain water for the whole year, so we find seven *beris* in its floor. When the water dried up by the beginning of summer, these *beris*/wells provided water. As the water body retains water for more than six months, the resultant percolation ensures continuous water availability in the *beris*, which were dug in its bed.[88] Moreover, subsequently, the dry bed of Amarsagar was used to cultivate *rabi* crops.

Tanks were among the other methods to harness rainwater. The construction of tanks required a thorough knowledge of the

[84] Chand, *Tawarikh Jaisalmer*, pp. 49–51.

[85] Chand, *Tawarikh Jaisalmer*, p. 41; Bhati, *Jaisalmer ri Khyat*, pp. 46–7 and 58.

[86] Here it is very important to recall the history of Jaisalmer. The ancient capital was situated at Lodurva and later on at Kuldara. However both these sites were short-lived, perhaps due to its inability to sustain a large urban population (see Plate 8 A & B).

[87] Misra, *Rajasthan ki Rajat Boonden*, pp. 49–50. However, with the growth of urbanization, the catchment area of these tanks is being encroached leading to gradual decay.

[88] Ibid., pp. 56–7.

PLATE 7A: GARSISAR, JAISALMER

PLATE 7B: PANORAMIC VIEW OF GARSISAR FROM JAISALMER FORT

geography of the area. The site was chosen carefully. The direction of the slope, its gradient and topography of the catchment area were important considerations. In India, there has been a very long

PLATE 8A: STEP WELL AT KULDHARA

PLATE 8B: INSIDE A STEP WELL

tradition of tank construction. The beginnings can be traced from the Great Bath of Harappa and water reservoir/s of Dholavira. It is interesting to note that Dholavira is surrounded almost from all sides by the salt water body forming the Rann of Kutch. Therefore, it was necessary to develop secure storage of potable water, especially to sustain a large urban population. The magnitude of the water reservoir of Dholavira leaves no doubts about the prevalence of technology to harness the rainwater and to construct a large tank with steps extending to the bottom of the tank.[89]

Behind the north gate, in the central zone of the citadel there has been exposed a 12.80 m wide water reservoir furnished with a 24 m long and 70 m broad inlet channel for carrying the rainwater which is so precious in that semi-arid environment. Its length remains indeterminate at present. The exposed part is only 9 m. Quite likely, it claims a large area of the citadel if the prevailing landscape thereabout is any indication[90] (see Plates 9 A, B & C).

The magnitude of the reservoir forces us to imagine the vitality of the reservoir as a supplier of water to the habitation. At the same time the magnitude of the reservoir also forces us to visualize the situation in the region if in any given year the monsoon fails; or to put it differently, the significance of the monsoon in semi-arid regions. It is difficult to define the agency which constructed the water reservoir at Dholavira due to our inability to read the script. Nevertheless, the tradition continued in the later periods also as we witness the construction of Sudarshan Lake. It is suggested that this lake was constructed under state patronage and subsequent epigraphic evidences make it amply clear. The tradition of construction of tanks received tremendous impetus during early medieval centuries, more prominently in peninsular India. 'The expansion of agrarian base and rural settlement region in Tamilnadu too was linked with the expansion of irrigation network.'[91]

[89] Bisht, 'Dholavira: New Horizons of the Indus Civilisation'.

[90] Ibid., p. 80.

[91] B.D. Chattopadhyaya: 'Introduction: The Making of Early Medieval India', in Chattopadhyaya, *The Making of Early Medieval India*, Oxford University Press, Delhi, 1994, p. 23.

PLATE 9A: DHOLAVIRA RESERVOIR WITH STEPS

PLATE 9B: WATER CHANNELS: DHOLAVIRA

PLATE 9C: CONCEALED WATER CHANNELS: DHOLAVIRA

As argued earlier also, R. Champakalakshmi has suggested that 'There is a general correspondence between the steady increase in irrigation works and the increase in the nadus under the Pallavas and early Cholas.'[92] In fact, one can argue that that the generalization made about Tamil Nadu would be applicable to other parts of peninsular India, though the pace and chronology of the formation of agrarian regions in such parts may have been somewhat different.[93]

A very important segment of water management in medieval times has been the provisions in urban centres usually centred in and around forts and palaces. Given the military strategic positions of most of the forts and palaces there was provision for sufficient water not only for regular use but also to manage affairs in case of a long siege. By definition most of the forts were located on hillocks. Such a location had an inherent disadvantage which made construction of traditional wells almost impossible. Thus rulers were forced to rely more on other sources of water than the ground water. We witness the construction of massive water bodies to harness the potential offered by other sources of water, primarily rainwater. A good example of water provisions in forts can be seen in the case of Ranthambore and Chittor; both are unfortunately no longer habituated. Correspondingly, good examples of urban water planning can be seen in the case of Jodhpur and Udaipur; both have remained continuously habituated since their inceptions.

Within the outer periphery of Ranthambore Fort one finds, along with several medium-sized and small water bodies, several prominent water bodies; Padam Talab, Jangali Talab, Suksagar Talab, and Kalasagar. These lakes are formed by natural inundations during the monsoon season and usually retain water throughout the year. These water bodies also help maintain the high water table in the adjoining areas. Jangali Talab is the biggest of these five water bodies.

[92] Champakalakshmi, 'The Study of Settlement'.

[93] Neeraj Sahay, 'Environment, Settlement History and the Emergence of An Agrarian Region in Early Medieval Andhra', unpublished Ph.D thesis, University of Delhi, 2009. Chapter 5 in particular deals with 'Agrarian Expansion and the Growth of Rural Economy', pp. 162–220.

Its *bund* is built in such a way that it obstructs the water flowing in from two different valleys. At the corner of the *bund* is a well from which water can be drawn for drinking. Similarly within the fort-palace there are several water bodies and most of them relied on the seasonal rainfall during the monsoon season. Ranee haud is situated near the palace and was probably the water source for the palace. It has a very large catchment area and some artificial canals to facilitate faster runoff. Ranee haud has a very long wall. . . . The Padma talab has a very high *bund* with flights of steps leading to the water and a pavilion. Its catchment is very large consisting of two different valleys. There is a well on the right side of the *bund*. . . . In Ranthambore Fort, water was for all practical uses, taken from wells next to *talabs* which recharged the wells. . . . All major structures in the fort are around these water bodies.[94]

All along the passage from the base on the outer periphery of the fort, one can still locate provisions for water for potable purposes both for humans as well as for draught animals; the sole source of draught power except humans. Similar provisions can be seen in the case of Chittor which perhaps due to similar factors remained almost impregnable, like Ranthambore. It is interesting to note that in both cases when forts were finally captured, victorious forces made it mandatory as part of the peace treaty that no repair can be made to the fort[95] (see Plate 10).

Chittorgarh Fort is also called the Water Fort. The fort had eighty-four water bodies, out of which only about twenty-two exist today. The fort spreads over an area of 700 hectares, 40 per cent of which is given over to water bodies. These include *talabs* (ponds), *kunds* (wells), and *baories* (step wells). All the *talabs* have a natural catchment. The average reservoir depth is about 2 m. Taken together, this means these reservoirs can store about 4 billion litres of water. The *kunds* and *baoris* are located below the *talabs*, so that even the seepage from the latter is not lost. In a year of more than normal rainfall (average annual rainfall: 700 mm), enough water would be stored to last the next twelve months. Even after water loss due to seepage and evaporation and other causes,

[94] Agarwal and Narain, *Dying Wisdom*, p. 158.

[95] Rima Hooja, *A History of Rajasthan*, Rupa and Company, Delhi, 2006, pp. 308 and 612.

PLATE 10: PROVISION FOR WATER: CHITTOR

an army of 50,000 could live in the fort for four years without fear of thirst. One of the large water bodies within the fort is known as Chitrang Talab. It is said that this *talab* was constructed during the reign of Chitrangad Maurya. The adjoining hilly terrain of the forts provides the ideal catchment for this water body. It usually retains water throughout the year and can be easily accessed by the draught animals also. Another important water body within the fort is known as Jaimal Fatta ka Talab. It also depends on the catchment from the adjoining areas and provided a backdrop to the Padmini Palace. Kukadeshwar Kund is another important water body located on a slightly lower ground than the Mahadev Kund. It is said that in AD 755 King Kukadeshwar got it built along with a temple of Kukadeshwar Mahadev on its banks. It is interesting to note that the seepage and sub-terrain water from the Mahadev Talab form an important source of water for the Kukadeshwar Kund. Another important water body is known as Gaumukh Reservoir. It is a deep tank filled by a spring coming from a 'cow mouth', situated at the edge of the cliff. It is considered to be sacred, where you can feed the fishes. This reservoir receives seepage from the Hathi Kund. There are several step wells, and tanks, also within the fort; Khatan Baori, Ghee-Tel Ki Baori, etc.; tanks and step wells are usually located below the large water bodies to harvest the seepage. Thus, even if the large water body dries up, the ground water available beneath the surface could still be harvested. For example, Surya Kund is located below Kalanad; Khatan step well is located below Fatehji ka Talab; and Bhimlat Kund is situated below Sukadiya Talab.[96]

A classic example of provision for water in the fort can be seen in the case of Jodhpur which continues to be habituated since its inception in the fifteenth century. It has been extensively documented by the Centre for Science and Environment.

> The city of Jodhpur was founded in 1495 AD. While selecting the location of the city, its rulers must have seriously considered its water potential and strategic situation. The Chonka-Daijar plateau is an important physical feature of the region. It is 30 km long, with a maximum breadth of about 5 km at the centre and a height of 120–150 m above ground level. . . . The whole plateau serves as the water catchment for 50 functional water bodies like *nadis, talabs*, tanks, and lakes, and indirectly for about 155 groundwater bodies like wells, *baories,* and *jhalaras*. The surface water bodies are primarily natural but they have been improved by the local people over the centuries. . . . Ancient surface water-bodies have been serving the local human and livestock population and wildlife for several centuries. . . . Each water body was provided with an extensive catchment, watercourses and canals to trap the precious rainwater.[97]

An extensive survey by the School of Desert Science (SDS) in 1989 was able to document the existence of most of these water bodies and has also highlighted the linkages among the water bodies, if any. The SDS has been able to locate 40 tanks and of these eight are located inside the city. The oldest is Ranisar constructed in AD 1460. Three *talabs*, namely, Chankelao (AD 1460) Phoolalao (AD 1490) and Naya Talab (AD 1700) have been lost. '. . . today the only safe *talabs* inside the city boundary are Ranisar and Padamsar (AD 1520). These *talabs* are the largest in size and they

[96] Agarwal and Narain, *Dying Wisdom*, p. 156.

[97] Ibid., pp. 115–17.

have maintained their identity. Water from these *talabs* is still used for drinking and other purposes by the local population.'[98] Most of these *talabs* have their own catchment and to a large extent are interconnected by a network of canals among themselves and with other water bodies.

> Jodhpur is perhaps the only city in the country where an all out effort was made to conserve every drop of rainwater. To achieve this, every catchment and hillock was drained by canals. Not only did each lake, tank and *talab* have an efficient canal system, there were interconnections between lakes and tanks to distribute water to all parts of the city.[99]

The cattle/herds required pastures, the natural vegetation for most part of the arid region. Yet, we come across the social practice of keeping common land accessible for animals near the village. This had the state's approval as well. This was known as *jor* (जोड)[100] The location of *jor* was chosen very carefully. It was usually located near the drainage area of the village so that waste-water could be used for irrigating the *jor*, thereby ensuring continuous water supply sustaining good pastures. These were usually located near the village.[101]

Jor were usually treated as common property owned by the village society at large. Nobody was allowed to encroach upon this land. However, evidence from the Amber region suggests that these were owned by the state. Any unauthorized use of its pasture was punishable.[102] *Pala* grass was extensively grown for pasturage purposes.[103] Nainsi informs us about a very big *jor* near the city of Pokaran[104] (see Plate 11 A & B).

[98] Ibid., pp. 111–19.

[99] Ibid., pp. 121–5.

[100] Nainsi, *Marwar ra Pargana ri Vigat*, vol. I, pp. 448, 452, 457, 461, 514, 516; vol. II, pp. 328 and 342.

[101] Ibid., vol. I, pp. 457 and 527.

[102] *Arshatta, Pargana* Bahatri, 1777 vs and *Pargana*, Naraina, 1791 vs, Historical Section, Jaipur Records, Rajasthan State Archives, Bikaner. Henceforth, HS, JR, RSAB.

[103] Nainsi, *Marwar ra Pargana ri Vigat*, vol. I, p. 515.

[104] Ibid., vol. II, p. 315.

PLATE 11A: *JOR*, BIKANER

PLATE 11B: *JOR* – OPEN ACCESS FOR CATTLE

VAGARIES OF MONSOON, IRRIGATION AND AGRICULTURE

In the absence of any perennial river, agriculture in the region depended upon rains. The close relationship between the rains and agriculture becomes more obvious in the following popular sayings:

डाल को चूक्यो बानरो ।
आषाढ़ को चूक्यो करसो ।।[105]

If a peasant fails to sow his field in the month of Ashadh (the first month of rains) when it rains, then he is like the monkey whose hands have slipped off the branch of the tree.

Similar concerns are echoed in other sayings:

आटियो पोस
कुण को दोस ।[106]

If you start your ploughing and sowing as late as in the month of Paus, then you should not blame anybody, as there will be no production.

The popular sayings were a medium to impart knowledge. The climatic features necessitated initiation of cultivation by a particular month and any delay would result in crop failure. Thus conventional wisdom is visible in these sayings.

The kind of crops to be sown was dependent upon the timing of the rains. If the rains got delayed, they would sow coarser crops like *moth*. Crops like *bajra, jowar* required frequent watering, whereas *moth* can sustain itself on two to three watering only. Pundit Harasram informs us that *bajra* (a *kharif* crop) did not require any more rains, whereas *moth, jawar* needed one or more shower.[107] A popular saying complements this.

[105] Govind Lal Joshi, 'Harauti Ki Krishi Kahavaten', *Maru Bharati,* vol. XIII, 1965–66, p. 44.

[106] Ibid., p. 47.

[107] *Arzdasht,* Asoj Vadi 11, 1756 vs, HS, JR, RSAB.

जौ़ कहे म्हारे माथे भालो, बाण पाण में रहूं आलो ।
एक पाण चूक जाउं तो बका दूं घर को टालो ।।[108]

Jaun (barley) needs regular and timely water supply, otherwise it will perish.

Agricultural production is a complicated and specialized function. It needs knowledge of not only the environmental features of the land but also the crops to be grown.

Knowledge is critical element in farming and brings all the elements together. Many texts therefore implicitly describe territories of social power in agriculture. Ideas moving among farmers create territories of knowledge. An elusive geography of ideas surrounds farmers who need to know how to make the best (or even safest and simplest) gamble with the rains. Each farmer needs to know about soil preparation, seed selection, planting, watering, manuring, and weeding for the specific combination of water, crops, soil, and labour conditions on each farm. Ways of knowing come from generations of learning in wide regions.[109]

The prerequisite characteristics for better crop production—water, number of ploughs and quantity of water required by different crops—have also been depicted in the contemporary sources.

The time required for the maturing of various crops varied. This was important because the gestation period for each crop differed. *Moth* needs the least time. Vakil Amar Chand and Sahib Ram informed the state that due to meagre and delayed rains, only *moth* could be sown in this kharif season. It took place even as late as Bhadava.[110] Ajitdas and Man Ram inform us that due to drought, not even *moth* could be sown,[111] resulting in a famine-like situation. These examples explain the uncertainties and resultant anxieties inherent in the agricultural production. The sheer statement that '*not even moth could be grown*' or '*moth is being sown*', explains the

[108] Joshi, 'Harauti Ki Krishi Kahavaten', p. 47.

[109] Ludden, *An Agrarian History*, p. 38.

[110] *Arzdasht*, Bhadva Vadi 7, 1774 vs and *Arzdasht*, Sawan Sudi 9, 1752 vs, HS, JR, RSAB. Traditionally Bhadva is the third month of the rainy season and usually the last one also.

[111] *Arzdasht*, Falgun Sudi 11, 1751, vs, HS, JR, RSAB.

total failure of rains in the season. This evidence also points out that *moth* could be grown very late.

The centrality of water to the human settlements and agriculture is beyond doubt, however, duration, quantity and quality of water made a significant difference to the agricultural practices. As pointed out earlier, Rajasthan has a very diversified character which is also evident in the nature of agricultural production. Most of the south-eastern region has supported two crops a year. *Kharif* cultivation depended almost entirely on the rains; intensive cultivation during the *rabi* and variety of lucrative crops were largely determined by the extent of availability of irrigation facilities. Multiplicity of crops was an important feature of the agriculture of the eastern region. In the *kharif* harvest the dominant crops were *bajra*, *moth*, *mung*, *urad*, *jowar*, etc. As such *bajra* cultivation dominated the region. Apart from food crops, cash crops like cotton, sugar cane, indigo, etc., were also grown. The range of crops grown during the *rabi* harvest was limited. Wheat, barley and gram dominated the *rabi* harvest. The more arid western and north-western part has generally sustained a single crop a year. However, with the help of irrigation facilities possibilities of two crops were always a reality. Similar to the eastern part, the cultivation of bajra dominated the crop in *kharif* and wheat in *rabi.*

The long-term and the short-term variations in the climate and rainfall could ruin the crops and dry up pasturage, thereby making its sustenance difficult. For instance, in AD 1717, it was reported that in *qasba* Aaveri, rain measuring ten fingers fell on Sawan *sudi* '11', so ploughing could be done only for four-five days.[112] Similarly, for pargana Amber in AD 1705, it was reported by *purohit* Harsram, that from Sawan *vadi* '7' to Sawan *sudi* '1', there was little rainfall in the pargana and because of strong winds, ploughing was not done and production of crops fell sharply.[113]

The practice of inter-mixing of crops was quite prevalent in the whole region. Numerous small millets were grown along with cereals with the object of replenishing the productivity of the soil;

[112] *Arzdasht*, Sawan Sudi 11, 1659 vs, HS, JR, RSAB.

[113] *Arzdasht*, Sawan Sudi 1, 1647 vs, HS, JR, RSAB.

the most important being *kangri*, *kodra*, *kuri*, *malicha* and *sama*. The creeping pulses *moong*, *urad*, *moth* were sown sometimes with *jowar*. The winter pulses *masur* and *tur* were grown to perform the same function.[114] Similarly, crops were rotated to conserve the productivity of the land. The system of crop rotation was based on the results of local experience.[115] For example, a cotton field was left fallow in the ensuing harvest when it was sown with maize in the autumn, barley in the following spring, maize again in the next autumn, after which it was left fallow during the spring before cotton was again sown in the autumn.

It will be wrong to suggest that agricultural practices and the nature of production has remained the same over the centuries. Along with social developments in terms of technological interventions and socio-political support for agrarian production through harnessing the irrigation potential, we come across several adaptations to the climatic variability. In the absence of any snow-fed river on the one hand and the erratic nature of rainfall on the other, the only viable alternative was to tap water through artificial methods of irrigation. People had evolved numerous indigenous methods to overcome this constraint. Along with several types of wells mentioned in the first section, *chanch*, *kosita*, *dhimra*[116] or *dhibra*, *kohar*, *kuwa* and *bera*, there were several methods of water harvesting for irrigation purposes like *khareen, rela, sewanj, bahla,* etc.

The most prevalent practice of irrigation has been the diversion of rainwater to the fields for irrigation purposes. During rains, the rate of downpour being very high, water tends to run off and if not checked goes waste. Rainwater was stored by directing it to the fields. One of the systems where rainwater was diverted, in case of surplus, to neighbouring fields was termed as *bahlo*. The land thus inundated was also used for cultivation of crops. In village Sanwarla of pargana Seewano, the drainage caused by rains on

[114] S.P. Malhotra, 'Man and the Desert', in Rakesh Hooja and Rajendra Joshi, (eds.), *Desert Drought and Development*, Rawat Publishers, Jaipur, 1999, p. 27.

[115] *Imperial Gazetteer of India*, vol. V, p. 150.

[116] Nainsi, *Marwar ra Pargana ri Vigat*, vol. V, p. 132.

the nearby hill was stored in the field and used for irrigation.[117] Village Motisaro of pargana Seewano could produce 200 *mun* of Sewanj wheat with the help of *bahlo* of the village.[118] In the village Mahawan of pargana Pokaran, the overflow of a *talab* was stored in a small tank nearby.[119] In village Kharia Neeba Ro of pargana Sojat, the land was fertile due to availability of run-off rains from the nearby hill.[120] Similarly, in village Peepalija Kapaalid Ro of pargana Jaitaran, the overflow was used both for agricultural purposes and to irrigate the *jor*—the village pastureland.[121] On these lands, crops could be sown; especially the ones which needed limited water.

Rela, a regional variation of *bahla* functioned on the same principle;[122] *bahla* utilized rainwater, whereas *rela* used overflow from seasonal streams. Such a method, however, could not be used for irrigating fields other than those inundated by the floodwater of the seasonal streams. Lands irrigated by the *rela* method were generally located along the banks of rivers.[123] These were generally natural depressions and put into productive use by the village communities. The *rela* sustained *rabi/unali* crops. The village Peepaliyo Kapadiya Ro utilized the *rela* method of irrigation to produce revenue amounting to Rs.1,091 for the year 1658 AD, Rs.2,250 for the year 1659, Rs.817 for the year 1662 AD and so on.[124] The villages Anandpur Khas and Lambia of *pargana* Merta were producing good quantity of wheat and gram due to availability of *rela* irrigated land.[125] It should be kept in mind that *rela* was the natural water spread over fields and the crop grown on this field was known as *sewanj*,[126] and these were *rabi* crops. Usually wheat was grown due to low requirement of moisture and

117 Ibid., vol. II, p. 238.
118 Ibid., p. 239.
119 Ibid., p. 329.
120 Ibid., vol. I, p. 427.
121 Ibid., p. 516.
122 Ibid., pp. 207, 454, 455, 515; Nvol. II, pp. 145, 239, 329, 346.
123 Ibid., vol. I, p. 353.
124 Ibid., p. 516.
125 Ibid., vol. II, p. 120.
126 Ibid.

limited watering. Generally winter showers provided the watering required for the better growth of these crops.

Another method of rainwater harvesting for irrigational purposes was *khareen* and practised in Jaisalmer region.[127] The details have already been discussed in earlier chapters.

At the same time, tanks and reservoirs too played an important role in irrigation. An incidental reference in the *Akbarnama* gives clear evidence of how the lakes and tanks were used for irrigation. When Akbar was at Merta, he ordered a reservoir previously used for irrigation to be cleaned.[128] The environmental features of the region necessitated tank irrigation. Lakes and dams played an important role as sources of irrigation in the Mewat region also.[129]

The rulers and nobles had constructed numerous small and large reservoirs. Pokaran and Phalodi were prominent for having a number of tanks. The reason may be ascribed to geographic features. The water table being very low and the water being brackish, tank construction was the only viable alternative. Incidentally, there were 659 tanks recorded for 521 villages in Marwar. The highest number of tanks were in Sojat, followed by Jaitaran and Siwana.[130]

For Marwar region, village-wise numbers of tanks have been provided by Nainsi and are presented in Table 4.2.[131]

These systems of irrigation help in the cultivation of either of the two crops, *kharif* or *rabi* or both. These were important mechanisms to sustain the cultivation in the region where land was in abundance and the major barrier was non-availability of water. The words *kharif* and *rabi* were scarcely known in Rajasthan; the autumn harvest was called *sialu*[132] and the spring *unalu*.[133]

[127] Chand, *Tawarikh Jaisalmer*, p. 92.

[128] Abul Fazl, *The Akbarnama*, vol. III, tr. H. Beveridge, Rare Books, Delhi, 1973, pp. 308–9.

[129] Mughal Emperor Babur extensively describes the Kotla Lake and noticed its utility for irrigation. It lies partly in the district Nuh, partly in Gurgaon, where the two tracts join at the foot of the Alwar hills. Babur, *Babur Nama*, p. 580.

[130] Bhadani, *Peasants, Artisans and Entrepreneurs*, p. 52.

[131] Ibid.

[132] Chand, *Tawarikh Jaisalmer*, pp. 211–12; Lalas, *Rajasthani Sabad Kos*. Any kind of literature in Rajasthan primarily use this term only.

[133] Chand, *Tawarikh Jaisalmer*, pp. 211–12.

TABLE 4.2: TANKS IN MARWAR REGION

Name of the Pargana	Villages where tanks recorded	No. of tanks	Average no. of tanks per village
Jodhpur	11	36	3.27
Sojat	157	172	1.09
Jaitaran	75	94	1.25
Siwana	84	96	1.14
Jalore	59	76	1.29
Merta	16	16	1.00
Sanchor	58	69	1.19
Pokaran	34	57	1.68
Phalodi	27	43	1.59

The former was more important because it covered a larger area. However, on the other hand, the money value of the spring harvest (*rabi*) was generally greater. It was natural for a region without any perennial rivers to depend solely upon the rains. Thus, for the economically lower rung of society, the *kharif* crop was the major supply of food grains. It has been estimated that the proportion of out-turn of food grains from *sialu* to that from *unalu* in the hill tract was eight to three, and in the open country, three to two. Crops, which needed irrigation, were classified as *piyal*.

Barley probably covered the largest area during the *unalu* crop. *Bajra* and *moth* were by far the most important food staples of the whole region.[134] Even when *rabi* crops were grown, the autumn harvest was far more relied on for annual food supply.[135] A nineteenth-century estimate puts the ratio of the output of food grains from *kharif* to *rabi* as roughly 9:1 in Bikaner and 11:5 in Marwar.[136] From our sources it is possible to ascertain which harvest predominated in which pargana, based on the revenue yield from them. Table 4.3 summarizes the findings of B.L. Bhadani.

[134] Tod, *Annals and Antiquities*, vol. II, p. 126; Boileau, *Personal Narrative*, p. 174; Bhadani, *Peasants, Artisans and Entrepreneurs*, pp. 62–4.

[135] Devra, 'Desertification and Problem', p. 212.

[136] Erskine, *Rajputana Gazetteers*, vol. 2-A, pp. 101 and 343.

TABLE 4.3: RELATIVE IMPORTANCE OF *RABI* AND *KHARIF* HARVESTS[137]

Name of the Pargana	Revenue realization (% of the total in 1664)	
	Rabi crop	*Kharif* crop
Merta	37.83	62.16
Sojat	54.93	45.07
Jaitaran	47.88	52.12
Siwana	30.57	69.42
Jodhpur	16.06	83.93
Phalodi	0.00	100.00
Sanchor	40.04	59.96

In the first category come the three parganas, Merta, Jodhpur and Siwana, and in the second the parganas Sojat and Jaitaran. In the parganas of the first category the percentage of realization from the kharif was substantially higher than that from rabi. The relative size of revenue contributed by the two harvests in the parganas of the first sort is as follows: In Merta, 37.83 per cent in rabi, 62.10 per cent in kharif, in Siwana 30.57 per cent, in rabi, 69.42 per cent in *kharif* and in Jodhpur 16.06 per cent in *rabi* and 83.93 per cent in *kharif* and in Sachor 40.04 per cent in *rabi* and 59.90 per cent in *kharif*. The pargana Phalodi does not record *rabi* at all. These figures broadly correspond to the nature of soil and availability of natural/artificial irrigation as discussed earlier.

To further strengthen our premise that the environment to a great extent influenced the subsistence pattern, especially agricultural production, Table 4.4 prepared by B.L. Bhadani is of immense help. This table contains the total number of villages of Marwar in which crops are recorded and the percentage of such villages to the total number of villages are provided.

The table makes it clear that Pokaran/Phalodi had a lower percentage of villages in which crops are recorded. This was the same region which was particularly deficient in artificial irrigation, as discussed earlier.

137 Bhadani, *Peasants, Artisans and Entrepreneurs*, p. 163.

TABLE 4.4: CROP INFORMATION COVERAGE IN NAINSI'S *VIGAT*[138]

Name of the Pargana	A Total no. of villages	B No. of villages in which crops are recorded	B as % of A
Merta	382	213	55.76
Jaitaram	142	89	60.13
Sojat	213	141	66.20
Siwana	118	78	60.10
Phalodi	59	23	38.98
Pokaran	76	20	26.31

In the agrarian economy of this semi-arid and arid region, the failure of monsoons very often led to droughts. These droughts could easily have assumed the form of a famine. Famines hit the material base of the peasantry crucially. The high cost of seed and manure and dry wells restricted irrigation. Ploughing could not be done[139] thus affecting the agricultural production in the region. The drought or infliction of crops, scarcity of fodder and dry ponds made cattle vulnerable,[140] affecting the pastoral communities of the region. Apart from expecting relief from the state, peasants tried to minimize the impact of drought by adapting various solutions. For example, if rains were delayed, peasants shifted to the cultivation of *moth* as it has the shortest maturity period. Kanwar Pal and Bhopat Ram inform the *sarcar* on Kati Sudi 9, AD 1698, that in the regions which have received insufficient rains, land is being worked upon to make it soft so as to sow the seeds.[141] Kanwar Pal and Hari Valabh Das supplement the information that wherever soil is hard, it needs one more shower and wherever rains have been insufficient, artificial irrigation through wells was arranged as well.[142]

138 Ibid., p. 65.
139 *Arzdasht*, Paus Sudi, 4, 1647 vs, HS, JR, RSAB.
140 *Sanad Parwana Bahi*, Falgun Vadi 10, 1672 vs, Jodhpur Records, RSAB.
141 *Arzdasht*, Kartik Sudi, 9, 1641 vs, HS, JR, RSAB.
142 *Arzdasht*, Asoj Vadi 7, 1642 vs, HS, JR, RSAB.

The dispute over the use of water for irrigation purposes in the semi-arid climate cannot be ruled out and we follow cases of arbitration by the state. It was reported from village Dhamorki, pargana Khori, that the *patel* had diverted water from the river without permission. He was punished with a fine of Rs.20.[143] Rupa Patel of village Raitoli breached the embankment of the river to divert water.[144] In another incidence, the state punished a Gujar for diversion of water from another person's field in *qasba* Baswa, pargana Bahatri in 1774 vs.[145] A similar dispute is reported from village Neembla, *pargana* Bahatri. Goverdhan Brahman fought with Tulani Brahman over the issue of diversion of water and hit him with an axe. He paid Rs.11 as fine.[146] It is to be noted that the village ponds were the common village resources. They supplied water for various uses. The village pond had been used not only for washing clothes and ancillary activities but also by cattle for drinking purposes. Thus, any act of its pollution was punishable. We have evidence from qasba Malpura, pargana Malpura that Lona dyer was fined Rs.5.50 for washing dyed cloth in the pond.[147]

The elite extended support for the construction of wells to encourage the habitation in the region to enlarge the revenue base of the states. Kings actively promoted the construction of mechanisms for water management. The construction of wells was promoted in the regions with a relatively high water table.[148] The state extended loans to the peasantry to excavate new wells.[149] Regions with deep water tables opted for tanks,[150] which were also state financed. Rulers also offered financial support for the repair of wells. Those peasants who dug up new wells were also offered

[143] *Arhsatta, Pargana* Khori, 1801 vs, HS. JR. RSAB.

[144] *Arhsatta, Pargana* Dausa, 1825 vs, HS. JR. RSAB.

[145] *Arhsatta, Pargana* Bahatri, 1774 vs, HS. JR. RSAB.

[146] Ibid.

[147] *Arhsatta, Pargana* Malpura, 1791 vs, HS. JR. RSAB.

[148] Bhadani, *Peasants, Artisans and Entrepreneurs*, p. 49. Wherever wells occurred more frequently in *c.* 1660, the water table was nearer the surface.

[149] *Arzdasht*, Kartik Sudi 13, 1686 vs; Bhadva Sudi 11, 1688 vs; Ashadh Sudi 15, 1688 vs; Sawan Sudi 13, 1685 vs and Asoj Vadi 12, 1695 vs, HS, JR, RSAB.

[150] Nainsi, *Munhot Nainsi ri Khyat*, vol. II, p. 52 and Das, *Vir Vinod*, vol. I, pp. 117–18.

concessions in the land revenue.[151] In pargana Pahari the failure of rabi crops has been attributed to the lack of well irrigation in most of the villages.[152] Similarly, the state was urged to provide loans to peasants in order to enable them to buy *lao-charas* (rope and leather) so that they could operate wells to irrigate the crops, which were being ruined.[153]

The governing class extended assistance by providing loans known as *taqavi*, which were recovered gradually with the growth of agricultural production.[154] Moneylenders extended loans to the peasantry at very high rates of interest and the peasantry had to accept it as there was no other option. The peasants could obtain a loan from the *mahajan* either against the mortgage of their property or against *malzamini* (surety) furnished by the village zamindar, patel or *jagirdar*.[155] However, the state imposed several restrictions on the functioning of the *bohras*. The general principle in regard to all loans—*tagai* as well as loans advanced by the *bohras*—was that debt incurred in the current harvest period should be recovered immediately after the harvest or within the fiscal year. Arrears on the older loans could only be demanded if the output of the current harvest was considered adequate. The state's attempt to discourage the recovery of old loans by the *bohras* suggests that such loans were generally unrecoverable and therefore written off.[156] It was perhaps also due to the consideration that excessive exploitation of peasantry would compel them to migrate and thus, ruin the production potential.

Rulers not only arbitrated on the issue of ownership of means of irrigation but also deliberated on the amount of water to be distributed.[157]At times the state had to interfere to decide the

[151] *Arzdasht*, Kartik Vadi 3, 1742 vs, HS, JR, RSAB.

[152] *Yaddashti pargana Pahari Mauza- Darobastki*, 1784 vs, Historical Section, Jaipur Records, Rajasthan State Archives, Bikaner.

[153] *Arzdasht*, Asoj Vadi 12, 1752 vs; Asoj Vadi 1, 1751 vs; Kartik Vadi 3, 1751 vs, HS, JR, RSAB.

[154] Habib, *The Agrarian System*, p. 295.

[155] Singh, 'The Role of Mahajans', p. 22.

[156] Ibid.

[157] *Sanad Parwana Bahi*, Asad Sudi, 8, 1708 vs, Jodhpur Records, RSAB.

amount of water due to various claimants from an irrigational device.[158] It is interesting to note that the state used to increase the land to be cultivated after the digging of new wells in the villages.[159] The state's concern for optimal use of irrigational facilities is also evident in the same document. Similar evidence is also available from Amber. An *arzdasht* written by Fateh Chand Ramji on Paus Vadi 5, 1773 vs, informs the king Sawai Jaisingh that well water is being used for irrigation as directed.[160]

Water harvesting has been an integral part of the expansion of settlements and growth of agriculture in this region. It is difficult to give a precise idea of the expansion. But an approximate study has been attempted here to locate a process of gradual but continuous appropriation of water management techniques and systems to offer better facilities for potable and irrigation purposes. A great amount of exchange of ideas and techniques have taken place between the region under scrutiny and nearby well-settled regions of the Indo-Gangetic doab in the north-east and Gujarat and Malwa in the south and south-eastern directions. It will not be out of place to mention that commercial interactions with the well-endowed regions on the east as well as west facilitated the process of state-formation in this region. Though difficult to map at the moment, but even technological exchange cannot be ruled out. The exchange of ideas and practices within the larger process of cultural integration played an important role in the better appropriation of the given landscape, technological knowhow and agrarian practices.

158 *Sanad Parwana Bahi,* Sawan Sudi 8, 1708 vs, Jodhpur Records, RSAB.

159 *Sanad Parwana Bahi,* Bhadva Vadi 7, 1708 vs, Jodhpur Records, RSAB.

160 *Arzdasht,* Paus Vadi 5, 1706 vs, HS. JR. RSAB.

CHAPTER 5

Visibly Invisible
Ecological Imprints

Apparently all-pervasive, but the social manifestations of ecological imprints usually remain invisible. A closer examination of the social fabric of Rajasthan will enable us to decipher the ecological imprint in the different walks of life. As discussed in earlier chapters, the monsoon-dependent ecologies of semi-arid and arid parts of Rajasthan usually practiced a form of agro-pastoralism. Generally, the pastoral component was very significant in the arid parts whereas the semi-arid part relied more on the agrarian segment in the broader agro-pastoral economy. This difference in the centrality of different components of the agro-pastoral economy in different ecological niches is a clear testimony of the influences of the environment. Within the larger agro-pastoral economy of the region a closer look at the ways and means of appropriation of surplus by different sections of the society will help us to decipher the diverse ecological imprints on different sections of the society. In other words, social stratification also shapes the possibilities of resource utilization. The distribution of resources was to a great extent conditioned by the social structure defined in terms of heirarchy and stratification. Similarly, the possibilities of production also influence the social structure. This two-way very subtle interaction has been dealt with in the first part of the chapter. The latter part of this chapter illustrates the ecological imprints on social practices, rituals, etc.

Traditionally, as illustrated in previous chapters, studies of the pre-colonial history of Rajasthan have focused largely on the glories of the martial tribe/s on the one hand and the agrarian system on the other.[1] As has been the tradition, most of these accounts treat the environment as a 'given entity', with little inferences drawn for historical processes. The dynamics of the environment in the day-to-day functioning of the society have been at the margins of these studies. These works have rarely contextualized the environmental features for a nuanced reading of the social realities and have unfortunately supported a tacit acceptance of the stereotypical images of pre-colonial societies. A deeper examination of social systems in early medieval times to a great extent has been restricted by the nature and paucity of historical sources. Of late historians have examined the ways in which the process of emergence and growth of settlements gradually culminated in the crystallization of regional polity.[2] These works have mostly relied upon epigraphic evidences, at times supplemented by the literature of the period. But, most of the scholars working on this period have generally highlighted the limitations of sources. The paucity of sources for the reconstruction of the social structure and/or of agricultural production has been very aptly lamented by Chattopadhyaya. He argues that:

> ... any idea of progress can be empirically substantiated only if sufficient comparable material is available for the early (Medieval) period, which, as we have seen, is not the case. Evidence on crop production in early medieval period has also to be strenuously culled from the mostly indirect information that the inscriptions offer.[3]

Nevertheless, with the growing access of historians to village-level administrative documentation of the seventeenth and eighteenth centuries, aspects hitherto little explored have come within the ambit of research. These documents have given deeper insights into the ecological imprints on political, economic and social

[1] Ojha, *Rajputane ka Pracheen Itihas*; Tod, *Annals and Antiquities*.

[2] Chattopadhyaya, *Aspects of Rural Settlements*; 'Origins of the Rajputs'; Kapur, *State Formation in Rajasthan*.

[3] Chattopadhyaya, 'Irrigation in Early Medieval Rajasthan'.

aspects of the period. Usually marginalized, literary evidences have enriched these narratives to some extent.[4] Thus we are in a position to critically evaluate the nature and functioning of social system/s within the larger ecological context.

AGRO-PASTORAL LANDSCAPE: SOCIO-POLITICAL NEGOTIATIONS

In pre-colonial Rajasthan, 'what is beyond dispute is that the political and social order were closely intertwined, integral parts of a continuum, webbed together through intricate ties'.[5] Society was characterized by the hierarchical order based on birth and the duties and privileges attached thereto.[6] This socio-political stratification determined, among other things, access to natural resources of the different sections of society. It is pertinent to point out at the very outset that unlike colonial states, in pre-colonial

[4] Along with archival records, these accounts have relied on *khyat* literature as well. Usually, literary sources derive content from the immediate socio-ecological settings; they have extensively been used to decipher the functioning of village society. Satish Chandra began an examination of the inner dynamics of village social structure with respect to revenue appropriation and distribution. 'Some Aspects of the Village Society in Northern India During the 18th Century: The Position and Role of the *Khud-Kashta* and *Pahi Kashta*', *Indian Historical Review*, vol. I, no. 1, 1974, pp. 51–64; Gupta, *The Agrarian System of Eastern Rajasthan*; Singh, *The State, Landlords and Peasants*, further developed it. They have examined the agrarian structure with reference to the state and social stratification, namely, the caste system. Scholars in such works as Sharma, *Rajput Polity*; Devra, *Rajasthan ki Prashashnik Vyavastha*; Sharma, *Rajasthan Through the Ages*; Bhadani, *Peasants, Artisans and Entrepreneurs* and others, to mention a few, have examined the issue of agrarian relations to investigate the nature of medieval Rajput polity. Their central focus has been on revenue realization and related issues. Recently, Nandita Prasad Sahai, *Politics of Patronage and Protest: The State, Society, and Artisans in Early Modern Rajasthan*, Oxford University Press, Delhi, 2006 has examined another very important but often neglected component of social structure, that is, artisans.

[5] Sahai, *Politics of Patronage and Protest*, p. 235.

[6] Dipankar Gupta, 'Hierarchy and Difference: An Introduction', in Dipankar Gupta (ed.), *Social Stratification*, Oxford University Press, Delhi, 1991, p. 141.

times the polity and kinship systems were coterminous, especially in Rajasthan.[7] At the top in the hierarchy was the head of the clan, who had acquired territorial rights and established the principality; then came the *thikanadars* (nobles) who were his *saga* (blood relatives), brothers, sons, etc. The dominant sections of the village society comprised *jagirdars*, *bhomias* and *bohras*, who owed their status partly to hereditary superior proprietary rights in land and partly to their position in the revenue administration.[8] *Mahajans/bohras* were the providers of borrowed capital for the ordinary peasantry. The rulers offered a certain amount of patronage and protection to *mahajans*, as the political powers recognized their importance.[9] Although the roles of various castes of society were defined by the *varna* system, the vagaries of monsoon-based agricultural production placed a certain level of flexibility on the role and functions of different sections of society.

The political apparatus ensured privileges not only for the wealthy and influential sections, but also for the weaker sections of society. Ritual prescriptions were also at times bypassed because of the need for sufficient supply of labour to work the cultivable land available in abundance.[10] This was not a condition peculiar to Rajasthan. The key to successful statecraft was typically the ability to attract and hold a substantial, productive population within a reasonable radius of the court. Given the relative sparseness of the population and the ease of physical flight, the control of arable land was pointless unless there was a population to work it.[11]

[7] Sharma, *Rajput Polity*, pp. 1–58.

[8] Gupta, *The Agrarian System of Eastern Rajasthan*, pp. 134–43; Singh, *The State, Landlords and Peasants*, pp. 42–50.

[9] Singh, 'The Role of Mahajans', pp. 20–31.

[10] Sahai, *Politics of Patronage and Protest*, p. 127, also refer to fn. 12 on the same page.

[11] James C. Scott, *Weapons of the Weak: Everyday Forms of Peasant Resistance*, Yale University Press, New Haven and London, 1985, p. 185; Sumit Guha has argued that Southeast Asia has always seen greater population density since ancient times. Sumit Guha, *Health and Population in South Asia from the Earliest Times to the Present*, Permanent Black, Ranikhet, 2001.

NEGOTIATING AGRO-PASTORAL LANDSCAPE

Political power cannot be measured solely in terms of geographical expanse. 'A growing, productive population settled in the ambit of a monarch's capital was a more reliable indicator of a kingdom's power than its physical extent';[12] this seems to be closer to reality. A continuous attempt on the part of the political apparatus to increase the number of settlements in their territory was clearly visible.[13] For the same reason migrations during the years of environmental distress were actively contained by offering monetary assistance and fiscal concessions along with material support.[14] Precisely for this objective, even peasants who had migrated were brought back with promises of concessions and remittances in revenue demand by the socio-political elite.[15]

The general geophysical features of the region suggest that there was inadequate availability of water resources leading to limited possibilities of human settlement and agriculture.[16] References in official documents of such settlements which perhaps did not contribute significantly to the state exchequer, and their methods of water appropriation are, on the one hand, testimony to the concerns of the state and, on the other hand also point out the level of constructive utilization of the landscape attained by the state and society to harness the monsoon rains. The nature of political authorities, based on the intricate web of socio-political relations made it possible to understand the landscape at the local

[12] Locating the relative strength of a state between occidental and oriental regions, Scott has stressed that Southeast Asia has traditionally witnessed low population density and therefore, the state actively guarded its human wealth. Scott, *Weapons of the Weak*, p. 185.

[13] *Arzdasht*, Jeth Sudi 4, 1742 vs; Falgun Sudi 2, 1743 vs and Jeth Sudi 4, 1742 vs, HS. JR, RSAB.

[14] The pre-colonial kingdom thus rode a narrow path between a level of taxes and coercive exactions that would precipitate wholesale flight. Scott, *Weapons of the Weak*, p. 185.

[15] *Arzdasht* Magh Vadi 9, 1765, vs; Ashadh Sudi 2, 1762 vs; Jeth Vadi 7, 1752 vs, and Chait Vadi 3, 1752 vs, HS, JR, RSAB.

[16] 'The alternative route through Rajputana was in general sparsely inhabited. . .', Moreland, *India at the Death of Akbar*, p. 11.

level, that is, villages. Decentralization of political authority was integral to the political formations to facilitate prompt responses to the local ecological conditions. The nature of interventions made by the administrations with respect to not only the sites for the construction of water systems, but also to the methods of distribution and utilization clearly suggests that the landscape had already been figured out in great detail and in a precise manner. As has been argued in earlier chapters, most of the methods were practised at the individual level, while some required larger community participation as well as political support. Wherever feasible, the most important method to secure water was wells. With the variations in the depth of the water table, the structures and functioning of the wells also changed. A lower water table necessitated deployment of water-lifting devices.[17]

A thorough and insightful understanding of the landscape on the part of the locals is visible in the conceptualization of these water systems and their efforts to harness the given potential were amply supported by the contemporary socio-political agencies. The ruling elites actively supported the development of such systems at local levels due to economic exigencies as it claimed a share in the enhanced agrarian production. The following section examines these interventions as manifestations of socio-political concerns for environmental factors as well as the responses thereof.

Given the sole dependence on monsoon rains, we have enough evidence to conclusively argue that the state extended financial and material support for the construction and maintenance of water systems to appropriate monsoon rains. Most of the investment was made to retain cultivators and thus to ensure the continuance of the settlements and its agricultural production. The ruling elite extended support for the construction of water bodies to encourage habitation in the region so as to enlarge the revenue

[17] These devices differed according to the topography and the depth of the water table. The selection of a water-lifting device differed according to the nature of requirement, that is, potable or irrigation. The state was urged to provide loans to peasants in order to enable them to buy *lao-charas* (rope and leather) so that they could operate wells to irrigate crops. *Arzdasht*, Asoj Vadi 12, 1752 vs; Asoj Vadi 1, 1751 vs; Kartik Vadi 3, 1751 vs, HS, JR, RSAB.

for the principalities. The system of mapping the physical environment clearly manifested in the nature of interventions made by the political agencies to mitigate the vagaries of nature in general and monsoons in particular is very perceptive. The administrative apparatus kept a strict vigil over human settlements and continuously explored the favourable landscape to develop new possibilities for expanding habitation. The ruling elites continuously aspired to make the landscape more congenial and cultivable and with the emergence and consolidation of Mughal power in north India, we witness a definite step forward. The *Ain-i-Akbari*, the administrative manual compiled during the reign of Akbar by Abul Fazl, made a beginning in this direction. It was soon imitated by Munhot Nainsi in his *Marwar ra Pargana ri Vigat* for Marwar which seems to have been replicated in the *arhsatta* documents maintained by the Amber state.[18] These documents represent a kind of politico-administrative mechanism that laid ever-increasing stress on both the physical and social mapping of the realm.[19] The administrative reorganization of territory

[18] This trend received unprecedented impetus with the beginning of colonial rule, which started collection of census data, compilation of gazetteers, and conducted anthropological studies from the nineteenth century.

[19] At another level, this kind of documentation was seen in European societies as part of their process of modernization which was visible at the level of abstract as well as realpolitik. It was the product of two processes working in synchronicity; first as Scott puts it, 'the process of documentation is best conceived as a strong, one might even say muscle bound, version of the self-confidence about scientific and technical progress, the expansion of production, the growing satisfaction of human needs, the mastery of nature (including the human nature), and above all, the rational design of social order commensurate with the scientific understanding of natural laws. It originated of course, in the west, as a by product of unprecedented progress in science and industry.' Second, in terms of 'utilitarian discourse which replaces the term "nature" with the term "natural resources", focusing on those aspects of nature that can be appropriated for human use. A comparable logic extracts from a more generalized natural world those flora and fauna that are of utilitarian value (usually marketable commodities) and, in turn, reclassifies those species that compete with prey on, or otherwise diminish the yields of the valued species. Thus, plants that are valued become "crops", the species that compete with them are stigmatized as "weeds", and the insects that ingest them are stigmatized as "pests". Thus, trees that are valued become

under Mughal rule, that is, the creation of *subah*, *sarkar* and *pargana* boundaries consolidated the administrative set-up of the principalities in this region.

Apparently similar, the administrative structures developed in Rajasthan differed broadly from the Mughal counterparts. In Rajasthan, the administrative apparatus was an extension of the larger social structure where historically power was shared among clan members. However, the efforts directed at strengthening the administrative structure may be seen as a political strategy to negotiate the diverse social[20] and ecological factors.[21]

At the hub of the revenue administrative set up was a *diwan*, who supervised the assessment and collection of land revenue from the *parganas*, each of which had its own officials. He had to keep himself well acquainted with the affairs of the principality, particularly with the activities of *pargana* officials such as the *amil*, *amin*, *chaudhari*, *qanungo* and *potadar*. They were his medium

---

"timber", while species that compete with them become "trash" trees or "underbush". The same logic applies to fauna. Highly valued animals become "game" or "livestock", while those animals that compete with or prey upon them become "predators" or "varmints"." J.C. Scott, *Seeing Like a State: How Certain Schemes to Improve the Human Condition Have Failed*, Yale University Press, New Haven, 1998, pp. 4–13. However, we do not find any similarity of purpose between these two distinct experiences.

[20] The nobles were being gradually but systematically converted into mere revenue assignees. See Dilbagh Singh, 'Regulating the Domestic: Notes on the Pre-colonial State and the Family', *Studies in History*, vol. XIX, no. 1, n.s, 2003, pp. 69–86.

[21] Water was the most determining factor for crop production. The supply of water by natural precipitation was meagre and sporadic, even while the climatically induced requirement for water was very incessant. Irrigation, thus by necessity, becomes the primary concern for crop production in the whole region. Means of irrigation have always been very important yet very limited. The availability of irrigation made a difference in agricultural productivity. The long- as well as short-term variations in the climate and rainfall ruined crops and dried up fodder, thus making living conditions difficult. In the absence of any snow-fed river on the one hand and the erratic nature of rainfall on the other, the only viable alternative for irrigation was to tap water through artificial methods of irrigation. Singh, *The State, Landlords and Peasants*, pp. 51–3 and Bhadani, *Peasants, Artisans and Entrepreneurs*, pp. 41–2.

of contact with village officials such as the *patel*, *muqaddam* and *patwari*. The *diwan* also ensured that none of these officials exacted illegal or forbidden cesses or such as had been remitted by the king.[22] The records maintained at the office of the *diwan* provided the administrative map and social profile, along with a synopsis of the seasonal variations pertaining to the ecological conditions.

The appointments of *pargana* officials such as the *amil*, *amin*, *potadar*, *chaudhari*, *qanungo*, and others were made by the *diwan*. The revenue collectors (*amils*) and assessors (*amins*) worked under the direct supervision and control of the *diwan*. He was required to see that all papers such as the rolls of *jama*,[23] *awarija*,[24] and *yaddasht*[25] were sent to him regularly. He issued instructions to the *pargana* officials for special measures regarding cultivation during the famine period.[26] Through such mechanisms, the polities of the region were able to read the variations in the ecological conditions and offered remedial measures.

As the *pargana* was a fiscal as well as a territorial unit, the *amil* and the *amin*, the land-revenue officers were responsible for the implementation of royal decrees and initiatives. The *amil* was entrusted with the task of ensuring cultivation on all arable land and the collection of the assessed revenue from the cultivated lands. He had to take the help of the *tappadar*, *chaudhari*, *qanungo*, *shahana* and other village officials. All letters of receipt and disbursement such as *arhsattas* had to be duly signed by the *amil* and the *amin*. The *amil* was also enjoined to expand the areas under cultivation by getting more land, especially *ujar* land (wasteland) brought under the plough. He was authorized to issue new *pattas* of *vat-vih* (share in the cultivation) at concessional rates, ranging from one-third to one-fourth to those cultivators who brought *parat* (fallow)

[22] Gupta, *The Agrarian System of Eastern Rajasthan*, p. 164.

[23] The amount of revenue assessed on an individual or a village or larger area.

[24] Literally, *awarija* is a diary, a daybook, a ledger, a rough notebook, an abstract of receipt and disbursement.

[25] A kind of *patta* granted to the *jagirdars* at the time of grant of *jagirs* by the rulers or at time of its restoration. Some of these have also recorded the number of ploughs and bullocks per cultivator along with their castes.

[26] Gupta, *The Agrarian System of Eastern Rajasthan*, p. 165.

lands under the plough. He regularly communicated information regarding famines and other natural calamities that affected the cultivation, and implemented the measures prescribed by the *diwan*, such as making provisions for *hal* (plough), *bij* (seed) and *bail* (oxen), for the *palti* (cultivators).

Instructions were also issued by the *amil* to the *chaudharis*, *qanungos*, *patels* and *patwaris* to search for and rehabilitate the *raiyat*, who had abandoned cultivation due to famine, among other reasons, and to provide them with the facilities for cultivation. The *amil* and the *amin* were both responsible for the collection of *baqaya* (arrears) in their *pargana*; the *diwan* demanded *muchalka* (paper of agreement) from the *amil* and the *amin* of each *pargana* owned responsibility for the realization of the arrears. The *amin*'s main duty was to verify the area of land under cultivation and to assess it according to the existing rules and regulations. He was required to send the assessment and other papers maintained in his office to the revenue department. He was also required to inform higher authorities about the condition of rains in his *pargana*. The complex web of administrative functionaries ensured greater and deeper intervention by the state in the rural landscape. This set up was the rulers' eye and ear.

Two other important *pargana* officials concerned with the land revenue administration were the *chaudhari* and the *qanungo*. The offices of the *chaudhari* and *qanungo*, though hereditary, were subject to renewal by the king. The hereditary nature of these offices further strengthens the contention that the political apparatus was an extension of kin relations. The *chaudhari* was supposed to render all help to the *amin* and the *amil* in the collection of revenue. He also served as a counter check against the *qanungo*. As regards the duties of the *qanungo*, he was required to collect detailed information about the land revenue and enter the same in the records under different heads. Both the *chaudhari* and the *qanungo* were required to prepare the *taqsim*[27] of different villages.

[27] *Taqsim* records carried the summary account of revenue in the *pargana* including its villages; the total area of a *pargana*, area of uncultivated land especially the area under *sir*, *kalar*, hill, *nullah*, etc. It also included area under

These records were very significant for the political agencies as they contained all the relevant details with respect to the socio-ecological settings of villages. They provided an insight into the agrarian potential of a village, taking into consideration the role of environmental factors.

The *shahana* was perhaps the lowest official in the hierarchy of the revenue administration. The main duty of a *shahana* was to keep a watch on the crops when they were ready, or when they were taken to the *kotha* (where the grain was stored for the purpose of sale) or to the *ganj* (grain market where it was brought for sale).[28]

Apart from the many officials appointed by the state, there were several others who were part of the political apparatus on account of their hereditary rights. Most of them were part of the local elite or the influential sections of society. Prominent among them was the *patel* who was normally a peasant (*raiyat*) and sometimes a petty zamindar. On the one hand, he was the natural leader of the villagers, and on the other he was the official headman (*mukhia*). He represented the *raiyat* and forwarded their grievances to the *pargana* officials. The *patel* was not paid directly from the treasury but he was entitled to some customary sources of income, such as the *inam muqaddami*, *bisondh* (one-twentieth part) and *dagli* (Rs.14 at the time of the foundation of a new village). He was given a *parwana* for *nankar*, a cash allowance or land in lieu of his services. Along with the *patwari* and the *shahana*, he would receive a coconut at the time of *batai* (sharing of the harvest).[29] The *patwari* was the village accountant who prepared the *khasra* papers as well as accounts of revenue collections. His papers were all in the local language. In the *khasra* papers, he was required to record the

---

cultivation, the area under *punya-udik* (charitable land) along with harvest-wise revenue realisation. Gupta, *The Agrarian System of Eastern Rajasthan*, p. 322.

28 Ibid., pp. 66–79.

29 A great deal of responsibility devolved upon the *patel* in times of famine. Alongwith the *quanungo*, it was his duty to rehabilitate the village, to bring *paltis* (cultivators) back to their fields, and to make provisions for seeds, bullocks and ploughs through loans either from the local *bohra* or from the *raj*. He was required to recommend grants of agricultural loans (*takai-taqavi*) by the state in times of distress. Ibid., pp. 183–4.

name of the *asami* (cultivator), the area of his holding (*arz* and *tul*), the cultivable area, the name of the crops sown in *sayallu* (*kharif*) and *unhalu* (*rabi*) and the share of the *raiyat* and the *diwan* (state) in both the categories (*khalisa* and *jagiri*) of land.[30] Thus, it can be safely argued that the state acquainted itself with various types of ecological facets and had penetrated deep into the intricacies of the environmental landscape for its politico-administrative needs.

A close look at the land revenue collection process and a reflection on agrarian production and output clearly establishes the correlation between investments made by the ruling elite on rainwater harvesting to improve the agricultural potential.[31] The differences in land revenue rates also reflect the same phenomenon. Irrigated land was taxed at a lower rate compared to the non-irrigated land, which is a good indicator of the state's concern to improve irrigation facilities. Local ecological as well as socio-political conditions provided ample opportunities to the political agencies to promote their agenda. While negotiating disputes over the use of water for irrigation purposes in the semi-arid climate the political powers had the platform to promote their own interests. Rulers not only arbitrated on the issue of ownership of means of irrigation but also issued directives on the amount of water to be distributed.[32] At times, the state had to arbitrate between claimants over the use of irrigational devices.[33] It is interesting to note that the political authorities ensured an increase in the land to be cultivated by digging new wells in the villages.[34] The state's concern for optimal use of irrigation facilities is also evident from this important document pertaining to *pargana* Sojhat of Marwar state. Similar evidence is also available from Amber principality. An *arzdasht* written by Fateh Chand Ramji from *pargana* Malarna dated Paus vadi 5, 1716 vs (AD 1659), informs Raja Sawai Jaisingh

[30] Ibid.

[31] Bhadani, *Peasants, Artisans and Entrepreneurs*, pp. 34–80. Dominance of *kharif* crops also reflects greater dependence on source of irrigation for the crops grown during non-monsoon/*rabi* season.

[32] *Sanad Parwana Bahi*, Sawan Sudi 2, 1765 vs, Jodhpur Records, RSAB.

[33] *Sanad Parwana Bahi*, Kartik Vadi 11, 1768 vs, Jodhpur Records, RSAB.

[34] Ibid.

that water from wells was being used for irrigation purposes as directed by the state.[35] Clearly, the state was aware of the ecological realities and therefore almost always attempted to maximize the utilization of available water.

An examination of the taxation system offers good insights on the constraints of the environment on social structure. While determining the magnitude of land revenue from any region, various factors were considered. The most important had been the immediate environmental features. Features like the quality of land, productivity of land, means and methods of irrigation used, kind of produce, etc., were considered to take care of limitations or possibilities.

There were two modes of revenue assessment prevalent in the region. These were, first, a cash assessment per unit area of land or the *zabti* system; and second, the crop-sharing method or *batai* based on a physical division of the crop into an agreed proportion between the peasant and the state. These two systems coexisted for each *qasba* and village and for each harvest.[36]

It is important to note that it was the type of crop grown which determined how the crop was to be assessed by the *zabti* or *batai* method or sometimes interchangeably by both. The crops usually assessed by the *zabti* system were the 'cash' crops in both *kharif* and *rabi* harvests. Cash crops were termed as *pukki* and were never assessed by the crop-sharing method. All the major food grain crops were usually assessed under the *batai* category. However, as these could be assessed under both the systems they were termed as *kutchi* indicating the flexibility. Whereas in an area of dry farming that was prone to indifferent harvests, crop sharing was clearly of advantage to the cultivator as the preferred mode of discharging his revenue obligations, we can say that in conditions of uncertain production, the preference for crop sharing, especially for rain-fed crops, is consistent with the risk-averse strategy of the peasant.

The fact that the assessment of revenue was based on the type of crop grown in the region rather than land being the criterion

35 *Arzdasht*, Paus Vadi 5, 1716 vs, HS, JR, RSAB.

36 Gupta, *The Agrarian System of Eastern Rajasthan*, pp. 39–44.

can be explained in terms of agronomic reasons. In the semi-arid and arid environment of Rajasthan, inter-cropping was extensively practised. Unlike mixed crops where crops are sown, harvested and sold as such, inter-cropped plants, whether or not sown together, are harvested separately. At the same time, the disparity in productivity and the prices led to the inter-cropped plants being assessed separately. In the case of such crop combinations, once each crop had been separately harvested, the physical division of the crops in the proportions specified by the *batai* schedule was perhaps the more appropriate and simpler method of revenue collection.[37]

It is important to point out that the cash crops were usually assessed under the *zabti* system. In *batai*, the risk of harvest failure was shared between the state and the peasant. However, loss on account of variations in local prices was borne by the state alone. In case of cash crops, the state might have attempted to negate the advantage of price variations to the peasant and thus, must have insisted on *zabti* rates. In the same context it is important to realize that growing cash crops required heavy capital investment in the form of manpower, irrigation, longer maturity period, etc. It implied that only the rich section of peasantry with sufficient resources could have grown cash crops. It once again strengthens our submission that the environmental features of Rajasthan must have intensified the social stratification.

It has been pointed out that some of the factors which determined the variations in the magnitude of land revenue demand are brought out in our documents. In practice, the difference in the quality of the soil, location of the field, duration of fallow of the field sown and the degree of capital and labour investment were incorporated to formulate a complex schedule of graded tax rates for each crop. Far from there being a uniform crop rate for a whole *pargana*, the rates even varied between fields within individual villages.[38]

[37] Madhvi Bajekal, 'Agricultural Production in Six Selected Qasbas of Eastern Rajasthan (*c.*1700–1780)', Unpublished Ph.D. thesis, University of London, London, 1990, pp. 111–15.

[38] Ibid., pp. 116–17.

In order to determine the productivity of land and its suitability for production, the land was classified into different categories. One of the more prevalent criterions has been the basis of irrigation facilities. The land was divided into two categories: irrigated and un-irrigated: *nadimatrka*[39] (river fed), *devamatrka* [40](rain-fed).

This classification highlights the significance of irrigation in the agricultural production of the region. Income from land revenue had been the mainstay of the state's treasury. Here one is reminded of the limited use of artificial irrigation some 200 years back. Shyamal Das says that both the crops depended upon rains for irrigation. The irrigation of land by the use of water from wells or from ponds was almost non-existent.[41]

Different rates of land revenue demand were sometimes prescribed on the basis of irrigation facilities. The tax burden on land artificially irrigated was less. It was practised perhaps to compensate the cost of its capital-intensive nature. For example, the crops cultivated in *bara* or *vor* or *varakyari* (land of the first quality) were charged at the rate of one-half, while the *piwal* or *piyal* (irrigated) lands watered by wells (*dhenkli*) and *tals* (tanks) were charged at a lower rate.[42] Land watered by river water was more heavily taxed per bigha (Rs.5) than that irrigated by wells (Rs.3) even when cash crops like sugar cane was produced.[43]

Similarly, the mixed wheat and gram crop produced on *sewanj* land was free from any extra charge; while in *piyal* wheat (wheat grown on land where land was irrigated by wells) 'official expenses' were levied at the rate of 1.5 *ser* per *man* (of produce).[44]

As discussed earlier, the rate of taxation varied according to the method of irrigation used. Variations in the magnitude of land revenue on account of this factor are most evident in the different rates for *kharif* and *rabi* crops. The magnitude of *mal* varied from

[39] Ibid., p. 118.

[40] Ibid.

[41] Das, *Vir Vinod*, vol. I, p. 118.

[42] Gupta, *The Agrarian System of Eastern Rajasthan*, p. 148.

[43] Bajekal, 'Agricultural Production in Six Selected Qasbas of Eastern Rajasthan', p. 118.

[44] Bhadani, *Peasants, Artisans and Entrepreneurs*, pp. 199–200.

pargana to *pargana* and from village to village within a *pargana*. It was higher in the *kharif* than in the *rabi* season, since the *kharif* crops were supported by rainfall and *rabi* required artificial irrigation. The average rate of *mal* in the *kharif* was 37 per cent and in the *rabi* about 30 per cent.[45]

Nainsi's *Vigat* provide details of *amal dastur* for *pargana* Merta. The *dastur* shows that the land revenue demand was imposed mainly in two forms, *batai* (crop-sharing) and *zabti* (cash revenue rates). It has been pointed out that 'under *batai*, land revenue in *pargana* Merta was collected at the rate of one-half (*bhog-adh-batai*) of the produce in *siyalu* (*Kharif*). . . . For the *unali* (*rabi*) the rate to be paid was 2/5 (*pachdui*) on the rain-irrigated land and one-third (33½) on the artificially irrigated (*piyal*) land sown with cereal crops.'[46]

The rate of taxation varied according to the nature of artificial irrigation devices as well. The demand was also fixed on the basis of means of irrigation. *Bhog* (land revenue) was, for example, stated as a rate on *arhat* (Persian wheels) in Jalore. The rate varies according to the kind of wells, as well as the castes of the owners.[47] This further testifies that the state was aware of the environmental limitations of the production process.

However, B.L. Bhadani has pointed out:

> In *unali* (*Rabi*), the tax-demand in *pargana* Pokran was made per *baori* (step well) at the rate of 1/3rd of the produce; but for the *sewanj* (mixed wheat and gram), barley and gram, the share was the same as fixed for cereals in *sawnu* (*kharif*) harvest. It implies that the total burden on the peasants was the same in *rabi* as in the *kharif*.[48]

There seems to be a problem in such an analysis. B.L. Bhadani has taken a different meaning of the term *sewanj* (mixed barley and gram), whereas it should have been un-irrigated land.[49] Thus

[45] Singh, *The State, Landlords and Peasants*, p. 111.

[46] Bhadani, *Peasants, Artisans and Entrepreneurs*, p. 199.

[47] Ibid., p. 210.

[48] Ibid., p. 201.

[49] Sita Ram Lalas, *Rajasthani Sabad Kosh*, vol. IV, pt. 3, Rajasthani Shodh Sansthan, Jodhpur, 1973, p. 5793.

we can say that constraints imposed by the environment on production possibilities in different parts of Rajasthan were recognized while determining the amount of land revenue.

In the state of Bikaner, the land was usually assessed at the rate of one-fourth of the produce or at Rs.4 per plough. As the state had very limited agricultural potential, so we find, instead of *bhog*, *rokad-rakam* as the main source of income from the countryside. It was as a nomenclature of those collective non-agricultural taxes which were always realized in *rokad* or cash as a part of land revenue. *Rokad-rakam* constituted a major part of the *hasil*. Generally it approximated 48 per cent or nearly 50 per cent of the total income of *hasil*.

Rokad-rakam consisted of a good number of taxes. The most important component was *dhuan bhachh*. It was a poll tax and was realised from each household at the rate of Rs.1. It was a major source of the *rokad-rakam* and contributed between 40 to 50 per cent of the total income. Similarly, a tax called *talibab* was levied on the non-agriculturist class at the rate of Rs.4 per family.[50] Whenever the state was scourged by famine, the ruler imposed an extra cess of 8 *annas* per camel load of goods brought by the merchants for sale in the state. The amount thus collected was spent for famine relief. It was known as *kal* tax.[51] The cultivators had to give one-fourth of the grass produced by them as tax.[52] *Rokad-rakam* was first introduced by Maharaja Anup Singh some time in AD 1668–69. It is interesting to note the justification extended for its imposition. By imposing the tax the Maharaja wanted to encourage the cultivators to enhance the fertility of their land and to undergo no extra burden on account of introduction of new taxes.[53]

The influences of the monsoon on agrarian production are one of the most visible aspects of human–nature interactions and have

[50] Powlett, *Gazetteer of Bikaner*, Appendix II, p. 162.

[51] *Kagad Bahi* 1827 vs, Bikaner Records, Rajasthan State Archives, Bikaner. Henceforth, BR. RSAB.

[52] *Kagad Bahi* 1839 vs, BR. RSAB.

[53] G.S.L. Devra, 'Nature and Incidence of Rokad Rakam (non-agricultural taxes) in the Land Revenue System of the Bikaner State (1650-1700 AD)', *Proceedings of Indian History Congress*, Calcutta, 1976, pp. 190-5.

The last resort of the peasantry to combat drought had been migration[63] to better regions. In a famine in AD 1665 the peasantry migrated towards Burhanpur, Kali Bhet and towards Agra and Mathura ji,[64] and in 1770 towards Malwa.[65] Migrations were also the result of peasant anxiety to save their cattle power in years of famine from any large-scale mortality. Since land was abundantly available, the possession of cattle was the crucial prerequisite for the process of cultivation and hence, peasants migrated with their cattle.[66] The state was actively involved in the rehabilitation of deserted villages.[67]

Indeed, preservation of the cattle wealth was the most important factor to explain migration in the western region. Since the cattle herder was also a peasant, migration was the only option.[68] The nomadic nature and a relatively low land : man ratio became the greatest asset in years of drought and scarcity. Herds were split up into smaller groups so as to disperse the animals over a wider grazing area. The natural vegetation of the western region suited the needs of cattle well. Mostly such migrations were a short-term phenomenon. In village Ranasar of Bikaner, due to famine in AD 1763 the *raiyat*, *chaudharis*, all migrated and settled in another village.[69] Similarly, in the year AD 1783 the village Lunkansar was totally deserted because of famine and not only the *raiyat* but also the *mahajans* and *chaudharis* left the area.[70] The option of migration was extensively used in the Marwar region also. In 1821 vs, the raiyat of village Laliya *pargana* Narnaul left the village and migrated.[71]

[63] *Arzdasht*, Asad Sudi 1, 1754 vs, HS, JR, RSAB.

[64] *Arzdasht*, Miti Sudi 2, 1722 vs and Asoj Vadi, 6, 1751 vs, HS, JR, RSAB.

[65] S.P. Gupta, and D. Singh, 'Famine in the Territories of Amber (1650–1770)', *Proceedings of Rajasthan History Congress*, Ajmer, 1975, pp. 52–5.

[66] *Arzdasht*, Asoj Sudi 2, 1762 vs, HS, JR, RSAB.

[67] *Kagad Bahi*, Kartik Vadi 3, 1840 vs, BR. RSAB; *Sanad Parwana Bahi*, Migsar Vadi 5, 1843 vs; Magh Vadi 9, 1821 vs, Jodhpur Records, RSAB.

[68] *Kagad Bahi*, Jeth Sudi 11, 1840 vs; Jeth Sudi 8, 1840 vs, BR RSAB.

[69] *Kagad Bahi*, Jeth Sudi 8, 1820 vs, BR RSAB.

[70] *Kagad Bahi*, Kartik Vadi 3, 1840 vs, BR RSAB.

[71] *Sanad Parwana Bahi*, Magh Vadi 9, 1821 vs, Jodhpur Records, RSAB.

Thus, it seems that natural calamities created problems for everybody, especially for the peasantry. The peasant relied solely on the stable growth of his annual crops. However, the above illustrations bring out various responses adopted by the peasantry. The peasant, being tied to his piece of land, did everything possible to save the crop; migration being the last option. However, the safety of cattle compelled them to move. We can say that in years of famine and scarcity, the range of options open to the pastoral communities was far wider than to the peasants.

The vagaries of nature and related uncertainties about the quantum of agricultural production necessitated flexibility in the methods of governance. Under such conditions, a kind of flexibility was necessary whereby revenue collection was not calculated in rigid terms, rather it was revised to correspond to the actual production generated. The possibility of cultivation of a particular crop in a season was determined by the timely onset of the rains and the choice of crop could vary according to the variations in the climatic patterns. Thus, flexibility had to be inbuilt in an agricultural economy where income or revenue differed from season to season due to cultivation of different crops with variance in yield also.[72] Considerable leeway was granted by the state to the local elite in the assessment of revenue. Revenue was collected on the actual agricultural production on the verification issued by the important elements and by consensus in local society.[73] On the testimonies of the local elite about the actual environmental conditions in both the agricultural seasons, various remissions were granted. It also strengthened the superior status of intermediaries. Intermediaries were as such part of the administrative apparatus and ensured two-way communication between the state and the peasantry.[74] These intermediaries safeguarded the interests of the state on the one hand and were the guardians of the peasantry on the other.[75] Intermediaries were dependent on the state for the

[72] Gupta, *The Agrarian System of Eastern Rajasthan*, pp. 38–73.

[73] Ibid., p. 164.

[74] *Arzdasht*, Magh Vadi 9, 1765 vs; Ashadh Sudi 2, 1762 vs; Jeth Vadi 7, 1752 vs and Chait Vadi 3, 1752 vs, HS, JR, RSAB; Habib, *The Agrarian System*, p. 295.

[75] Rohan D' Souza, 'Rigidity and the Affliction of Capitalist Property: Colonial

continuance of *pattas* while they needed the peasantry to cultivate the land, thereby ascertaining the smooth process of agrarian and ancillary production. From the central authority to the cultivator inherent flexibility can be observed with respect to agricultural production.[76]

Any discussion on the relationship between environmental features and society at large in Rajasthan is incomplete unless the popular perceptions behind drought and famine are analysed. Recurrent drought in this region usually led to famine. Colonel Tod calls it the grand natural disease of the western region.[77] The evidence from our region throws light on numerous strategies adopted by the common man to survive.

In Rajasthan famine was a regular feature. Famines of different kinds, namely, *annkal* (great famine), *jalkal* (scarcity of water), *tinkal* (scarcity of fodder) and *trikal* (scarcity of fodder, water and grain), have been a common occurrence in the region. The historical situation is revealed in a common traditional legend summarizing the expectation of periods of harsh conditions:

> *7 akal, 27 zmana, 63 kuria kacha and 3 akal arra pare maan poot ne Nile na Pachelia.*[78]

In a span of 100 years there would be seven famine years, 27 good years, 63 years of average harvest and three extreme famine years.

Any failure of monsoon would have led to drought. The relationship between rains and drought was very clear to the contemporary mind. It is quite evident in the various popular sayings described earlier.

Land Revenue and the Recasting of Nature', *Studies in History*, vol. XX, no. 2, n.s., 2004, p. 255. To demonstrate their superior social status and economic position, the ruling elite practised the euphemisation of economic resources. '. . . what the wealthy did was to transmute a portion of their disproportionate economic means into forms of status, prestige and social control by means of acts they passed off as voluntary acts of generosity or charity. This social control was, of course, again convertible into labour services and hence again into material wealth.' Scott, *Seeing Like a State*, pp. 306–7.

[76] Singh, 'The Role of Mahajans', p. 22.

[77] Tod, *Annals and Antiquities*, vol. II, p. 235.

[78] Malhotra, 'Man and the Desert', p. 17.

Along with various methods to combat drought and periods of scarcity, peasants turned towards religion. For example, when there were no rains till the *amavasya* of the month of Sawan, prayers were offered to Lord Shiva and food was distributed among Brahmins. It was reported that such efforts led to good rains.[79] We have already discussed that the works of charity were organized on behalf of the state as well. What is important here is that interpreting the immediate cause of drought and famine in terms of religion enabled the peasants to give environmental disturbance a meaning intelligible to themselves, while also enabling them to believe that by the performance of certain rites in a particular mode they could seek to end their affliction.[80]

EXPANDING THE REALM: APPROPRIATING FLORA AND FAUNA

In Rajasthan, especially the central and western parts, the vegetation was very thin and sparse. There were very few forests in the region. In such a situation it was necessary to protect the already existing ones. Lalchand complained to the Amber ruler on Jeth Vadi 1, 1756 vs about the tree felling in his *pargana* and expected punishment for the culprits. We have reference from Jaipur records about cash punishments for such crimes.[81]

The state was concerned about the protection of vegetation. Rulers took keen interest in the upkeep of gardens. The rulers of Amber were informed by Ganga Ram that gardens were being destroyed due to non-availability of water but they had been revived with the help of water from a canal. [82] In Jaisalmer region, a *bund*, Jaitsar,[83] was constructed to capture the runoff water from

[79] *Arzdasht*, Sawan Sudi 7, 1707 vs, HS, JR, RSAB.

[80] David Arnold, 'Famine in Peasant Consciousness and Action: Madras, 1876–8', in Ranajit Guha (ed.), *Subaltern Studies: Writings on South Asian History and Society*, vol. III, Oxford University Press, New Delhi, 1984, pp. 71-2.

[81] *Arhsatta, Pargana*, Sawai Jaipur, 1798 vs, HS, JR, RSAB.

[82] *Arzdasht*, Magh Vadi 2, 1773 vs, HS, JR, RSAB.

[83] Chand, *Tawarikh Jaisalmer*, pp. 49–51; Mishra, *Rajasthan ki Rajat Boonden*, pp. 55–6.

the northern hills. The reservoir created as a result of the bund fed the garden situated on the other side of the embankment. A small canal was also provided to carry the water from the *bund* to the garden. This reservoir could carry water sufficient for four–five months only. The dry bed of the reservoir was used to cultivate *unali/rabi* crop. Thus, it served several purposes.

The attitude towards tree conservation is reiterated in the following anecdote written by Nainsi in *Khyat*. He informs us that Maldevji got the *babool* trees of Merta cut. In response to it, Viram Deo said that he would cut the mango trees of Jodhpur. However, people advised him not to do so as trees were to be protected. Hence he chipped a small branch of Mango symbolizing that he has settled the account.[84] In this anecdote, the chief is restrained from cutting trees by his advisors and supported conservation. A very important example in this regard was the representation of the *khejari* tree in the official flag of the Bikaner kingdom. It can be ascribed to the accordance of due recognition to the natural vegetation of the region.

It is very difficult to compare these instances and practices of environmental conservation with the modern-day environmental movements. It is necessary to bear in mind that modern-day environmental movements are products of problems emerging due to industrialization and colonialism. However, in the context of medieval Rajasthan environmental conservation was a part of attempts made by the state or\and religious sects to protect the natural resource base.

The attempts made by states cannot be separated from their concerns to protect and maximize revenue from the natural resources. We have several evidences when punishment was awarded for cutting green trees. The punishment was always realized in cash and became a part of the state's revenue. In village Saithal, *pargana* Bahatri, a person was punished for cutting down a *neem* tree.[85] *Neem* has tremendous medicinal properties, and

[84] Nainsi, *Munhot Nainsi ri Khyat*, vol. III, p. 101.

[85] *Arhsatta*, Village, Kundala, *pargana* Bahatri, 1745 vs; Village Bilhata, *pargana* Bahatri, 1774 vs; Qasba Jaipur, *pargana* Jaipur, 1798 vs; Village Deewara *pargana* Malpura, 1774 vs; Village Manauli *pargana* Malpura, 1772 vs; Qasba

thus, needed protection. Being a medicinal plant, it was considered inauspicious to cut *neem*, thus the punishment. Similarly, cutting of trees of *peepal*,[86] and *bad*,[87] was considered inauspicious. At the same time, it is important to point out that *peepal* and *bad* had been worshipped; thus religious considerations also led to punishment. However, the punishment was substituted by cash payment. The fine became part of the state's revenue and religion had no role to play either in the collection or expenditure of this amount.

Alongside, we have evidences of punishments for cutting *jamun* trees,[88] etc. Cutting *babool* trees was also punished.[89] *Babool* had been part of the natural vegetation of Rajasthan and it needed little or no care. In the arid part, *babool* was the dominant tree and provided food for the camels. Considering the economic and ecological value of *babool*, it was necessary to punish people who tried to cut it.

It signifies that the state was not protecting any particular tree, rather it was interested in the extraction of maximum revenue and protecting its economic interests. In this context, it is important to point out that even cutting of grass from forests or hills was punished.[90] Moreover, reducing the practice of conservation to

---

Malpura *pargana* Malpura, 1772 vs; Village Bhabhuvas *pargana* Bahatri, 1819, vs, HS, JR. RSAB.

[86] *Arhsatta*, Chandpur, *pargana* Bahatri, 1775 vs; Village Mahin Nala Khurd, *pargana* Bahatri, 1781 vs; Village Brahmanvas, *pargana* Gazi ka Thana, 1807 vs; Village Mahwa, *pargana* Tonk, 1858 vs; Village Sairora *pargana* Malpura, 1821 vs; Qasba Baswa *pargana* Bahatri, 1722 vs; Village Maheba *pargana* Malpura, 1774 vs, HS, JR. RSAB.

[87] *Arhsatta*, Village Chauroti, *pargana* Hindaun, 1785 vs; Village Shivpur *pargana* Tonk, 1777 vs; Village Nuka Kivas *pargana* Lalsot, 1820 vs, HS, JR. RSAB.

[88] *Arhsatta*, Village Nadu *pargana* Bahatri, 1774 vs, HS, JR, RSAB.

[89] *Arhsatta*, Village Palasoli, *pargana* Naraina, 1791 vs; Qasba Baswa *pargana* Bahatri, 1745 vs; Village Jaisinghpura, *pargana* Bahatri, 1775 vs; Village Kalouta, *pargana* Bahatri, 1774 vs; Village Kankroli, *pargana* Bahatri, 1701 vs; Village Nadu, *pargana* Bahatri, 1701 vs; Village Madhubanpur, *pargana* Phagi, 1707 vs, HS, JR, RSAB.

[90] *Arhsatta*, *Pargana* Bahatri, 1777 vs; *Pargana*, Naraina, 1791 vs, HS, JR, RSAB.

mere economic explanation shall be only a partial explanation. The conservation of trees has been a very old practice and part of culture.

Furthermore, it is not only that tree conservation was practised in the society. The evidence from Amber principality highlights that the state even punished those who killed animals. The killing of cattle was punished.[91] We have pointed out the economic significance of cattle in the semi-arid and arid regions of Rajasthan. An *arzdasht* dated Jeth Sudi 14, 1854 vs informs us that when a cow was killed, the culprit was expected to return a cow of gold to the owner of the cow.[92] The significance of cows in the contemporary society can be realized from this evidence.

Pastoralism had been an important component of economy. At the same time, the usefulness of cattle in agricultural production cannot be neglected. Cattle had been part of rural households since the ancient past and an economic asset. Like trees, cattle protection has been part of culture and religious teachings also.

We have evidence that even the killing of a monkey was punished.[93] The protection being offered to animals has been pointed out in the context of religious sects. The killing of crocodiles was also punished.[94] Perhaps fishing was also prohibited or it might have been restricted to ensure regeneration for future consumption.[95]

[91] *Arhsatta*, Qasba Naraina, *pargana* Naraina, 1812 vs; Village Kiratpura, *pargana* Bahatri, 1774 vs; *pargana* Naraina, 1812 vs; village Kankaroli Kalan *pargana* Bahatri, 1774 vs; Village Dhawani *pargana* Sawai Jaipur, 1795 vs; Qasba Jaipur *pargana* Sawai Jaipur, 1798 vs; Village Harneri, *pargana* Gazi Ka Thana, 1807 vs; Village Mohamadpur, *pargana* Chatsu, 1775 vs, HS, JR, RSAB.

[92] *Arzdasht* Jeth Sudi 15, 1854 vs, HS, JR, RSAB.

[93] *Arhsatta*, Qasba Malpura, *pargana* Malpura, 1821 vs, HS, JR, RSAB.

[94] *Arhsatta*, Village Sakheena, *pargana* Tonk, 1858 vs, HS, JR, RSAB.

[95] *Arhsatta*, Village Vadhar, *pargana* Tonk, 1858 vs; Village Jamalpur, *pargana* Hindaun, 1785 vs; Village Talab, *pargana* Bahatri, 1786 vs, HS, JR, RSAB. It is general practice in India to not to eat fish during the rainy season as it broadly coincides with the regeneration period.

ECOLOGICAL IMPRINTS

It can be said that Rajasthan is a fantastic mosaic of agricultural and pastoral economy. The requirements of this remarkable mosaic on the region's resources have been exceedingly varied. The demands on natural resources were socially managed. These methods of resource utilization were dominated by the wider concerns of long-term sustenance. It was more so as they had very limited knowledge and control over various natural phenomena. The natural resources were available through agricultural production, pastoral activities, mineral resources, forest produce, etc. Society in different parts of Rajasthan had adapted to the production possibilities in their respective parts. As pointed out in previous chapters, the semi-arid nature of the environment had imposed several limitations on production possibilities in Rajasthan. Therefore social management of the environment in the form of cultural response was an important mechanism to manage the meagre resources. The social management also took care of the conservation of resources. To sustain agriculture as well as pastoral activities in the semi-arid and arid environment, monsoons and other environmental features were very important. Thus, it is relevant to analyse the popular understanding of various environmental phenomena.

The attitude towards the conservation of nature is apparent in the teachings of various sects like the Vishnois. Jasnathji also endorses such a viewpoint. Hira lal Maheshwari[96] has highlighted that Jambhoji had prohibited cutting green trees and animal slaughter for his followers. One plausible explanation is that the economy was primarily pastoral in nature. Hence any slaughter, even during droughts, would reduce the means of livelihood. Similarly, the cutting of green trees was prohibited, as it would reduce the availability of green fodder for the animals. It becomes more important in this region where natural vegetation was very thin and sparse.

[96] Hira Lal Maheswari, *Jambhoji, Bishnoi Sampradaya aur Sahitya*, pt. 1, B.R. Publications, Calcutta, 1970, pp. 434–57.

saint, Tejaji, also sacrificed his life to protect the livestock from invaders.[103]

At another level we notice that in any pre-modern agricultural society, rituals, social beliefs, superstitions and anthropomorphism occupy an important place. This was particularly true in the case of Rajasthan, which has a very inhospitable terrain. These rituals reflected human response to the local environmental conditions. They encompassed a wide range of human sentiments including fear, respect, appeasement and adoration.

Apparently, the most important social concern was availability of water through rains. We have numerous evidences where even the state had granted official sanctions for the performance of rituals and religious practices to ensure timely and good rains. An *arzdasht* writer Mota informs the ruler that as per the instruction *bad* worship was organized to appease the rain-god.[104] Similarly, in the year 1774 vs, *Arzdasht* written by Roop Chand and Ajay Ram informs that as per the directive of the ruler *brahma bhoj* had been organized on behalf of the state to ensure better monsoons.[105] The following *Arzdasht* of Roop Ram and Himmat Ram suggest that the year was pretty bad in terms of monsoon rains as even they were directed by the ruler to organize *brahma bhoj* and *sava lakh ka jap*.[106]

An important ingredient of culture has been celebration of festivals which reflect a close connection with the primary production cycle. The prominent festivals being celebrated in Rajasthan are Holi, Diwali, Teej, Nag Panchami, Makar Sankranti, Vasant, Gangore, etc. These festivals were primarily religious and had mythological or historical origins. It is very difficult to define the origin of festivals but broadly speaking, they were related to the different harvesting seasons. 'On the Amavus, or *ides* of Cartic, is one of the most brilliant fetes of Rajasthan, called the *Dewali*', says

[103] Pemaram, *The Religious Movements in Medieval Rajasthan*, p. 37.

[104] *Arzdasht*, Bhadva Vadi 5, 1762 vs, HS, JR, RSAB; Similarly, Nath Ram informs the state officials about *bad* worship. *Arzdasht* Bhadva Vadi 5, 1762 vs, HS, JR, RSAB.

[105] *Arzdasht*, Sawan Sudi 7, 1774 vs, HS, JR, RSAB.

[106] *Arzdasht*, Sawan Sudi 7, 1774 vs and Sawan Sudi 10, 1774 vs, HS, JR, RSAB.

Tod. Further he relates it with production season, '*Lacshmi,* though on this festival depicted under by the type of riches, is evidently the beneficent *Anapurna* in another garb, for the agricultural community place a corn-measure filled with grain and adorned with flowers as her representative.'[107]

Holi had been another important festival celebrated in the region.[108] It has also been identified with fertility/productivity. Regarding Holi, D.D. Kosambi opines that 'is a fertility orgy that may be traced back to the late stone-age. Late Mesolithic deposits of ashes, with a few animal bones (from the sacrifices), and rain-compacted strata proved annual or periodic recurrence in the locality associated with gigantic *holi* bonfires. Here is proof that the agrarian economy had won at last . . .'[109]

Tod has described Gangore in the following terms, clearly identifying it with the production cycles.

> . . . in honour of *Gouri*, or Isani, the goddess of abundance . . . when nature in these regions proximate to the tropic is in the full expanse of her charm, and the matronly Gouri casts her golden mantle over the beauties of the verdant Vassanti (personification of Spring). The meaning of *gouri* is "yellow", emblematic of the ripened harvest, when the votaries of the goddess adore her effigies, which are those of a matron painted the colour of ripe corn; . . . here she is only seen as Ana-purana, the benefactress of mankind',[110]

Although the interactions between humans and nature are very complex and most of the times they have become part of everyday lifestyle, these are difficult to identify clearly and lack of documentation further creates problems for better historical investigations. Most of the time, one has to grapple with the manifestations of influences exerted by nature on the humans and the other way round. There are still numerous manifestations of influences which could not be deciphered in this chapter and are

107 Tod, *Annals and Antiquities of Rajasthan*, vol. I, p. 475.

108 Ibid., pp. 452–3.

109 D.D. Kosambi, *An Introduction to the Study of Indian History*, Popular Book Depot, Bombay, 1975, p. 199.

110 Tod, *Annals and Antiquities of Rajasthan*, vol. I, pp. 454–5.

a reminder of the fact that in any complex society, it is difficult to segregate spheres of influence between the social and natural. None the less, it is interesting to note that the current research agenda has taken a lead in this matter and this will benefit history writing in general and the study of Indian historical traditions in particular.

CHAPTER 6

Epilogue

The Indian subcontinent like any other part of the globe has been witnessing climate variability in the historical past, though with varying intensity. The degree of variability is a very subjective term and adaptation to the variability has also been contingent upon the nature of social negotiation/s. A sort of consensus is emerging that the climate will witness higher levels of variability again, calling for immediate action and long term planning to mitigate its adverse effects. The developing world is devoting great amount of financial support to develop new understandings about the functioning of complex societies in a changing climate. They are emphatically trying to translate climate modelling and research into tools for managing complex agrarian systems in the face of climate change and other environmental risks.[1] Most of the climate modelling

[1] Decision Centre for Desert City-II Mission statement. Point No. 6. DCDC II is an interdisciplinary research centre advancing knowledge, education, and community outreach for urban climate adaptation, supported by National Science Foundation, USA. The National Science Foundation has funded a group of Decision Making Under Uncertainty (DMUU) collaborative groups for the past five years. The goals of DMUU collaborative groups have been to support research, education, and outreach that increase basic understanding of decision-making processes and of the information needed by decision makers; to develop tools to support decision makers and increase their ability to make sound decisions; and to facilitate interaction among researchers and decision makers. In addition, NSF's Human and Social Dynamics priority area supported

to a great extent is based on patterns of climate in the past. History of climate is being relied more and more to understand the future climate patterns. In the largely rural setting of Indian society, economy and settlement patterns, traditional methods of adaptations still hold good, though growing lack of flexibility is a cause of concern. Traditional methods of water harvesting are being increasingly incorporated in the contemporary urban housing. The contestations over water will accentuate with long-term variability in climate. Imminent possibilities of climate variability make it imperative on the part of the contemporary societies to be flexible and elastic. Possible implications of climate variability make it essential on our part to document past processes of adaptations and draw solutions, if possible. Traditional method of inter-cropping against mono-cropping of contemporary times needs to be highlighted as it, to a great extent, took care of the rainfall variability. Traditional agrarian practices, prevalent at least till the beginning of the nineteenth century, had this inbuilt mechanism to mitigate inconsistency of monsoon rains to a great extent.

Significance of adaptations, to the monsoon in the emergence, development and sustenance of human civilization in India in general and in the arid and semi-arid landscape of Rajasthan in particular has once again been reiterated in the preceding pages. Locating the significance of monsoon rains in the agrarian and pastoral landscapes of Indian desert and adjoining areas has been attempted on the basis of literary and archival sources. The all-pervasive character of monsoon makes it at times invisible to the historical documentation and thus it is difficult to gather pointed references to monsoon. However, that does not mean that there are no fingerprints of monsoon. A closer look at the literary, archival, archaeological, sources makes it amply clear that monsoon has left and still leaves prominent marks on the human society. In contemporary times the need to understand the finer footprints

interdisciplinary groups that addressed questions related to change and dynamics in human systems more broadly. Available on http://dcdc.asu.edu/about-us/reports/Accessed on 03-04-2012.

of monsoon has gained immense urgency. The section, 'Monsoon Landscape' has examined the cultural comprehension of monsoon in India along with scientific understanding. The way monsoon has been woven within the broader social fabric and ethos reinforces its significance for the past societies. Ever-growing writings on the monsoon, scientific as well as popular, are a clear testimony to the growing awareness towards the significance of monsoon. Similarly, greater appropriation of the traditional systems to harvest the monsoon rains are ample demonstration of the significance of historical investigations into the functioning of these systems.

Social appropriation of monsoon ecologies has numerous variants and most of them have shown tremendous amount of fluidity to adapt to the fluctuation inherent in these ecologies. The section on *Rain-catchers* has been a modest attempt on my part to capture the range of appropriations over temporal and spatial variations. An attempt has been made to document a transitional phase in the emergence and spread of human settlement and expansion of agrarian economies in the marginal areas. It has been argued that this proliferation of human settlements even in the marginal areas became possible with active exploitation of monsoon runoffs. Most of the times this was achieved by *in situ* appropriation and subsequently with greater acquaintance with methods, more complicated modes were developed. Gradual but continuous assimilation of *Devamatrka* along with *Nadimatrka* on the one hand signifies better comprehension and appropriation of monsoon rains but also simultaneously signifies penetration of agriculture in the areas not covered by possibilities of inundation agriculture. However, it will not be inappropriate to state that this expansion to a great extent provided the necessary backdrop which led to penetration of settled agriculture to the regions well beyond the immediate vicinities of rivers. In other words, it led to greater connectivity and integration of agrarian economies located along the rivers in the north, south and central India. Gradually one also notices elaborate decorative construction of step wells and water-tanks. It may perhaps be a reflection of the growing ritual significance of water-harnessing systems but it definitely testifies to a certain degree of prosperity, most probably due to enhanced agricultural production. Most of the times, the decorative motifs have been drawn from the larger Hindu pantheon of god and goddesses thereby attesting to the spread of agriculture under the broader umbrella of Hinduism at least in the early medieval period.

Nevertheless, what is the most interesting feature of agrarian expansion in the marginal areas has been the dependence on the pastoral ways of living as supplementary mode of sustenance. It will not be an exaggeration to suggest that the important reasons for the success to a transition from *Nadimatrka* to *Devamatrka* especially in the marginal areas has been the presence of pastoral ways of living to offer alternative sustenance in the event of failure of agricultural production in a season or couple of seasons. Significance of the pastoral component was also seen in the taxation system of states located in the arid parts of Rajasthan. Presence of pastoral components also signifies society's capability to be flexible and change location if need be. The pastoral component also offers supplementary sources of sustenance. Along with dietary support pastoral component could also ensure connectivity with other habitations around the place. An important characteristic of the pastoral communities has been the tendency to forge exchange networks and these networks proved vital for the sustenance of even the agrarian settlements of the region. Contrary to conventional portrayal that state in pre-colonial times was primarily concerned with the appropriation of agrarian surplus and did not invest in the growth of primary agrarian production. The monsoon ecologies of Rajasthan offer us a very different picture of state apparatus. Polities be it located in the semi-arid or arid parts of the region, display a very different kind of nature and character. Perhaps constraints of the ecology, gave a particular character to these polities. Vagaries of monsoon and general paucity of other sources perhaps compelled these states to actively engage with the ecological constraints and to secure a regular sustenance. At times, income from the agrarian surplus was supplemented through taxation on the pastoral communities. Such a practice was more prevalent in the arid sections of Rajasthan along with active encouragement to the traders to pass through their territory. Several concessions in transition duty were offered to the traders as this was an important source of revenue for the state. Safety and security of the traders was guaranteed and any disruption was severely dealt with. It was accepted responsibility of the local officials and assignees to ensure safe transition of the caravans.

Equally important has been the examination of different states' concern with respect to violations of the commons. The range of commons covered along with water resources to flora and fauna of the region. It is difficult to explain state's active interventions in punishing the violators in terms of a deliberate and well conceived environmental conservation policy. Moreover, it will be wrong to look for evidences of environmental conservation policy in the polities of pre-colonial times. Nevertheless, imposition of penalties on the culprits testifies to the reach of state administration to the deepest corner of the territory. Furthermore, it also challenges notions of local self-sufficient village economies and of village panchayats as units of local self-governance or at least a body to resolve local village level conflicts. The nature of punishment is an interesting field and needs further studies as most of the penalties were imposed in cash and were expected to be deposited in the treasury. There is no evidence that these penalties were to be utilized for the ecological concerns in the modern uses of the term. Another important consideration of such a practice of imposition of penalties in cash is a testimony of prevalence of cash economy in the region. The sheer fact that the common man could deposit penalties in cash is a clear testimony that states were actively practising and promoting money economies. It may have perhaps gained momentum due to integration within the larger Mughal economy. Further investigations in the field will definitely help us broaden our understanding of the character of the economy and will challenge the conventional imagery of self-sufficient economies where transactions in cash were not a regular phenomenon.

However, what is interesting is the evidence of deliberate attempts on the part of polities to offer protection to the deprived sections of the society whenever monsoon has been erratic and sustenance was challenged. Relative paucity of labour and greater climatic instability had led to premium on the available manpower. Numerous concessions and incentives offered to the peasantry to come back and continue with agricultural activities. There were several restrictions on moneylenders to recover the capital and interest beyond a point. State stood surety against the loans offered to the peasantry. What is interesting to note is the fact that state did not curb the practice of usury and it allowed moneylenders to continue with their occupation. It is perhaps a reflection of the limited capacity of the state to extend its reach beyond a point. Moneylenders were an important source of finance and at times even offered loans to the state itself. Moreover, they functioned as local treasury and offered facilities of *hundi*—money transfer from village to state treasury. The organic connectivity with regular give and take between different sections of society and mutuality of interdependence was tested and reinforced especially during the times of natural exigencies. Constraints for a section of society, socio-political as well as ecological, appear to have created opportunities and possibilities for another section of society. Monsoon variability on the hand, made the peasantry vulnerable and dependent upon the moneylender for financial support either in kind or in cash and on the other hand same ecological constraints made it possible for peasantry to extract greater concessions from the state in terms of rebate in taxes, etc.

Diverse ecological nichés and equally diversified corresponding social interface generated multiple understandings of different regions. Human social interactions with unique but within broader monsoon ecological settings of the region has often led to very diverse adaptations. These adaptations have been continuously evolving and responding to climate variability and numerous changes in technological intervention within the larger socio-politico-economic situations. Fortunately, late medieval period, generally in India and specifically in Rajasthan has witnessed intensive documentation of landscape, which perhaps is indicative of a deeper penetration by political apparatuses. Bayly has aptly pointed out, 'In other words, royal legitimacy was itself predicted on loyalty to and protection of homelands and bodies of subjects which were conceived as valued resources.'[2] The regional identities in relation to unique ecological conditions are very much visible in the emergence and growth of literature in regional languages. Though at a nascent stage, a notion of nationalism/proto-nationalism can be identified. Further studies are required to examine the possibilities of crystallization of nationalism/proto-nationalism based on ecological regionalism. Representation of *khejari* tree in the official flag of Bikaner kingdom is a clear example of how ecology has offered markers of identity formations. It can be ascribed to the accordance of due recognition to the natural vegetation of the region. A very significant example in this regard can be seen in the following remark:

> 17th century court chronicle Nainsi informs that, once territory between Hardas and Sekha was divided on the basis of grass it supported. Sekha took control of land, which produced *Karar*, while the Hardas got the *Bhurat* land. The above division very clearly points out that regions were identified by the kind of vegetation it supported. Even the kind of grass grown in any region became the basis of political identification of regions. Similarly, when Rao Jodha (ruler of Marwar) defeated Rana Kumbha (ruler of Mewar), it was decided that land, which sustained trees of *Babool*, would be bestowed to Marwar, whereas the *Aanwla* land would go to Mewar. Once again, the political division, as pointed out earlier, very clearly highlights the point that land was identified by the kind of vegetation it supported. The division of territory between regional polities on the basis of natural environmental features once again highlights the significance of environmental context. Although Rao Jodha was victorious yet, his acceptance of land dominated by the growth of *babool*—low productivity and arid conditions, as his territory and acceptance of claim of Rana Kumbha for the region with thick growth of *Aanwla*—more fertile region, substantiate the fact that regional polities stressed their identification on specific ecological settings.[3]

[2] C.A. Bayly, *Origins of Nationality in South Asia: Patriotism and Ethical Government in the Making of Modern India*, Delhi, 1998, p. 27.

[3] Mayank Kumar, 'Perceiving Landscape: Regional Identity Formations in Early Modern India', *History Today: Journal of History and Historical Archaeology*, no. 9, 2008, p. 59.

Bibliography

PRIMARY SOURCES

Archival

Arhsattas, AD 1689–1806 for *parganas* Pahari, Jalalpur, Piragpur, Sawai Jaipur, etc., Historical Section, Jaipur Records, Rajasthan State Archives, Bikaner.

Arzdasht, AD 1680–1720, particularly for the famine years AD 1684–6, 1690, 1696–7, 1704–5, 1712 and 1717, Historical Section, Jaipur Records, Rajasthan State Archives, Bikaner.

Kagad Bahi, VS 1820, 1821, 1826, 1840, Bikaner Records, Rajasthan State Archives, Bikaner.

Sanad Parwana Bahi, nos. 1, 2, 3, 44, 45, etc., Jodhpur Records, Rajasthan State Archives, Bikaner.

Yaddashti pargana Pahari Mauza-Darobastki, 1784 VS, Historical Section, Jaipur Records, Rajasthan State Archives, Bikaner.

Descriptive list of the Vakil reports addressed to the rulers of Jaipur, vol. II (Persian), Rajasthan State Archives Bikaner, Bikaner, 1972.

Epigraphic

Inscription of *Jhalarapatan* (Jhalawar) of the year VS 1143 (CE 1086), *Journal of Royal Asiatic Society of Bengal,* Calcutta, New Series, vol. X, no. 6, 1914, pp. 241–3.

Published Sources

Rajasthani

'Jaisalmer Ri Khyat', in *Parampara* (ed. Narain Singh Bhati), pts. 57–8, Rajasthani Shodh Sansthan, Jodhpur, 1981.

Lakhmi Chand, *Tawarikh Jaisalmer,* Rajasthani Granthagar, Jodhpur, 1899 (1999 rpt.).

Munhot Nainsi, *Marwar-ra-Pargana-ri-Vigat* (ed. Narain Singh Bhati), vol. I, Rajasthan Oriental Research Institute, Jodhpur, 1968.

———, *Marwar-ra-Pargana-ri-Vigat* (ed. Narain Singh Bhati), vol. II, Rajasthan Oriental Research Institute, Jodhpur, 1969.

———, *Marwar-ra-Pargana-ri-Vigat* (ed. Narain Singh Bhati), vol. III, Rajasthan Oriental Research Institute, Jodhpur, Jodhpur: 1974.

———, *Munhot Nainsi ri Khyat* (ed. Badri Prasad Sakaria), vol. I, Rajasthan Oriental Research Institute, Jodhpur, 1960.

———, *Munhot Nainsi ri Khyat* (ed. Badri Prasad Sakaria), vol. II, Rajasthan Oriental Research Institute, Jodhpur, 1962.

———, *Munhot Nainsi ri Khyat* (ed. Badri Prasad Sakaria), vol. III, Rajasthan Oriental Research Institute, Jodhpur, 1964.

———, *Munhot Nainsi ri Khyat* (ed. Badri Prasad Sakaria), vol. IV, Rajasthan Oriental Research Institute, Jodhpur, 1967.

Shyamal Das, *Vir-Vinod,* 2 vols., Motilal Banarsidass, Delhi, 1886.

Persian (Translation)

Abul Fazl, *Ain-i-Akbari*, vol. I, tr. H. Blochmann, Oriental Books Reprint Corporation, Delhi, 3rd edn., 1977.

———, *Ain-i-Akbari*, vol. II, tr. Col. H.S. Jarrett, Oriental Books Reprint Corporation, Delhi, 3rd edn., 1978.

———, *Ain-i-Akbari*, vol. III, tr. Col. H.S. Jarrett, Oriental Books Reprint Corporation, Delhi, 3rd edn., 1978.

———, *The Akbarnama*, vol. III, tr. H. Beveridge, Rare Books, Delhi, 1973.

Babur Zahirud Din Muhammad, *Baburnama,* tr. A.S. Beveridge, Oriental Books Reprint Corporation, Delhi, 1970

Badauni Abdul Qadir, *Muntakhab-ut-Twarikh*, vol. II, tr. W.H. Lowe, Academica Asiatica, Patna, 1973.

Jahangir, *Tuzuke-i-Jahangiri*, ed. H. Beveridge, Munshiram Manoharlal, Delhi, 1968.

European

Adams, Archibald, *The Western Rajputana States, A Medico-Topographical & General Account of Marwar, Sirohi & Jaisalmer*, Vintage Books, London, 1990 (1st pub. 1899).

Baden Powell, B.H., *The Indian Village Community with Special Reference to the Physical, Ethnographic and Historical Conditions of India*, Cosmo Publications, Delhi, 1972.

Boileau, A.H.E., *Personal Narrative of Tour through the Western States of Rajwara in 1835 Comprising Beekaner, Jesulmer & Jodhpur*, Baptist Mission Press, Calcutta, 1837.

Rennell, Major J., *Memoir of a Map of Hindustan*, W. Bulmer & Co., London, 1793.

Tavernier, Jean Baptiste, *Travels in India*, tr. V. Ball, Oriental Books Reprint Corporation, Delhi, 1977.

Temple, R.C. (ed.), *The Travels of Peter Mundy in Europe and Asia (1608–1667)*, vol. II, Hakluyt Society, London, 1914.

Tod, James, *Annals and Antiquities of Rajasthan*, vols. I and II, Rupa & Co., Delhi, 1997.

Administrative Records

Erskine, K.D., *Rajputana Gazetteers*, vol. II-A, *The Mewar Residency*, Scottish Mission Industries, Ajmer, 1908.

Imperial Gazetteer of India, vol. I, Clarendon Press, Oxford, 1907.

———, vol. V, Clarendon Press, London, 1909.

———, vol. VI, Clarendon Press, London, 1908.

———, vol. VIII, Clarendon Press, London, 1908.

———, vol. IX, Clarendon Press, London, 1908.

———, vol. XIII, Clarendon Press, London, 1908.

———, vol. XIV, Clarendon Press, London, 1908.

———, vol. XV, Clarendon Press, London, 1908

———, vol. XXI, Clarendon Press, London, 1909.

Letter no. 28, Letters Relating to the Protestant Mission to the East Indies, vols. 1-7 (SPCK), vol. 1, 1710–16.

Powlett P.W., *Gazetteer of the Bikaner State*, Office of the Super-intendent of Government Printing, Calcutta, 1874.

———, *Gazetteer of Ulwar*, Sallantyne Press, London, 1874.

The Rajputana Gazetteer, vol. I, Office of the Superintendent of Government Printing, Calcutta, 1879

SECONDARY WORKS

Acharya, R.M., *Sheep and Goat Breeds in India*, FAO Animal Production and Health Paper 30, Food and Agriculture Organisation of the United Nations, Rome, 1982. http://www.fao.org/docrep/004/x6532e/X6532E01.htm (accessed on 17 May 2010).

Agarwal, Anil and Sunita Narain (eds.), *Dying Wisdom: Rise, Fall and Potential of India's Traditional Water Harvesting Systems*, Centre for Science and Environment, Delhi, 1997.

Agarwal, Govind 'Rajasthan ke Jan-Jeevan aur Lok Sahitya main Oont', *Maru Bharti*, vol. XIII, 1965–6, pp. 11–29.

Aggarwal, P.C., *Caste Religion and Power: An Indian Case Study*, Shri Ram Centre for Industrial Relations, Delhi, 1971.

Agrawal, D.P., *The Copper Bronze Age in India*, Munshiram Manoharlal, Delhi, 1971.

Agrawal, R.C., 'Rajasthan Ke Katipaya Prachin Sarovar Tatha Vapi', *Varda*, vol. VI, no. 4, 1963, pp. 57-63.

Allchin, F.R., *The Archaeology of Early Historic South Asia: The Emergence of Cities and State*, Cambridge University Press, Cambridge, 1995.

Arnold, David and Ramchandra Guha, 'Introduction: Themes and Issues in the Environmental History of South Asia', in David Arnold and Ramchandra Guha (eds.), *Nature, Culture and Imperialism: Essays on the Environmental History of South Asia*, Oxford University Press, Delhi, 1995, pp. 1-12.

Arnold, David, 'Famine in Peasant Consciousness and Action: Madras, 1876–8', in Ranajit Guha (ed.), *Subaltern Studies: Writings on South Asian History and Society*, vol. III, Oxford University Press, Delhi, 1984, pp. 62–115.

——, *Famine: Social Crisis and Historical Change*, Basil Blackwell, Oxford, 1988.

Aymard, Maurice and Harbans Mukhia (eds.), *French Studies in History*, vol. I, *The Inheritance*, Orient Longman, Delhi, 1988 and vol. II, *The Departure*, Orient Longman, Delhi, 1990.

Bajekal, Madhvi, 'Agricultural Production in Six Selected Qasbas of Eastern Rajasthan (*c.*1700–1780)', unpublished Ph.D. thesis, University of London, London, 1990.

Bayly, C.A., *Origins of Nationality in South Asia: Patriotism and Ethical Government in the Making of Modern India*, Delhi, 1998.

Bhadani, B.L., 'Well-irrigation in Marwar in the 17th Century', *Shodh Patrika*, vol. XL, pt. 1, pp. 54–70.

——, *Peasants, Artisans and Entrepreneurs: Economy of Marwar in the 17th Century*, Rawat Publication, Jaipur, 1999.

——, *Water Harvesting, Conservation and Irrigation in Mewar, (AD 800-1700)*, Manohar, Delhi, 2012.

Bhadani, B.L. and Jibraeil, 'Technique of Irrigation in the Desert Area of Phalodi in Medieval Times,' *Proceedings of Rajasthan History Congress*, vol. XXIII, Bikaner Session, 2008, pp. 36-43.

Bhargava, Meena, 'Changing River Courses in North India: Calamities, Bounties, Strategies—Sixteenth to Early Nineteenth Centuries', *The Medieval History Journal*, vol. X, nos. 1&2, 2007, pp. 183–208.

Bhattacharya, Neeladri, 'Introduction', *Studies in History*, vol. XIV, no. 2, n.s. 1998, pp. 165-71.

Bhattarchaya, D.K., *Ecology and Social Formation in Ancient History*, K.P. Bagchi & Company, Delhi, 1990.

Bisht, R.S., 'Dholavira: New Horizons of the Indus Civilisation', *Puratattva*, vol. XX, 1989–90, pp. 71–82.

Bloch, Marc, *Feudal Society*, tr. L.A. Manyon, Routledge and Kegan Paul, London, 1961.

Botkin, Daniel B., *Discordant Harmonies: A New Ecology for the Twenty-First Century*, Oxford University Press, New York, 1990.

Braudel, Fernand, *Civilisation and Capitalism 15th–18th Century*, vol. III: *The Perspective of the World*, tr. Sian Reynolds, Collins, London, 1984.

——, *The Mediterranean and the Mediterranean World in the Age of Philip II*, vol. I, tr. Sian Reynolds, Collins, London, 1981; vol. II, tr. Sian Reynolds, Collins, London, 1982.

Brynt, Edwin F. and Laurie L. Patton (eds.), *The Indo-Aryan Controversy: Evidences and Inferences in Indian History*, Routledge, London and New York, 2005.

Bulliet, Richard W., *The Camel and the Wheel*, Harvard University Press, Massachusetts, 1975.

Burke, Peter (ed.), *A New Kind of History: From the Writings of Lucien Febvre*, tr. K. Folca, London, 1979.

Chakravarti, A., *Contradiction and Change: Emerging Patterns of Authority in a Rajasthan Village*, Oxford University Press, Delhi, 1975.

Chakravarti, Ranabir, 'Agricultural Technology in Early Medieval India (*c.* AD 500–1300), *The Medieval History Journal*, vol. XI, no. 2, 2008, pp. 229–58.

Chakravarti, Ranabir, 'The Creation and Expansion of Settlements and Management of Hydraulic Resources in Ancient India', in Richard

Grove, Vinita Damodaran and Satpal Sangwan (eds.), *Nature and the Environment*, Oxford University Press, Delhi, 1998, pp. 87–105.

Champakalakshmi, R., 'The Study of Settlement Patterns in the Cola Period: Some Perspectives', *Man and Environment*, vol. XIV, 1989, pp. 91–101.

Chandra, Satish, 'Some Aspects of the Village Society in Northern India During the 18th Century: The Position and Role of the *Khud-Kashta* and *Pahi Kashta*', *Indian Historical Review*, vol. I, no. 1, 1974, pp. 51–64.

——, 'Society, Culture and the State in the Medieval India', in Satish Chandra, *Historiography, Religion & State in Medieval India*, Macmillan, Delhi, 1996, pp. 193–235.

Chattopadhyaya, B.D., 'Early Memorial Stones of Rajasthan: A Preliminary Analysis of their Inscriptions,' in B.D. Chattopadhyaya, *The Making of Early Medieval India*, Oxford University Press, Delhi, 1994, pp. 120–9.

——, 'Introduction: The Making of Early Medieval India', in B.D. Chattopadhyaya, *The Making of Early Medieval India*, Oxford University Press, Delhi, 1994, pp. 1–37.

——, 'Political Processes and Structures of Polity in Early Medieval India,' in B.D. Chattopadhyaya, *The Making of Early Medieval India*, Oxford University Press, Delhi, 1994, pp. 183–222.

——, *Aspects of Rural Settlements and Rural Society in Early Medieval India*, K.P. Bagchi & Company, Calcutta and Delhi, 1990.

——, *The Making of Early Medieval India*, Oxford University Press, Delhi, 1994

——, 'Irrigation in Early Medieval Rajasthan', *Journal of the Economic and Social History of the Orient*, vol. XVI, pts. 2-3, December 1973, pp. 298-316.

——, 'Origins of the Rajputs: The Political, Economic and Social Processes in Early Medieval Rajasthan', *Indian Historical Review*, 3, 1976, pp. 59–82.

——, 'State and Economy in North India: Fourth Century to Twelfth Century', in Romila Thapar (ed.), *Recent Perspectives of Early Indian History*, Sangam Books, Delhi, 1995, pp. 309–46.

Chaudhuri, Tapan Ray and Irfan Habib, (eds.), *Cambridge Economic History of India*, vol. I, Cambridge University Press, 1982

Churnawat, Lakshmi Kumari, 'Bargawat Gatha: Itihas ke Jharokhon Se', in Jawahar Lal Handoo (ed.), *Folklore of Rajasthan*, Central Institute of Indian Languages, Mysore, 1983, pp. 75–94.

Cipolla, Carlo M., *Before the Industrial Revolution: European Society and Economy, 1000–1700*, Methuen & Co., London, 2nd edn., 1981.

Cohen, Bernard I., *Revolution in Science*, The Belknap Press of Harvard University Press, Cambridge, Massachusetts, London, 1985.

D' Souza, Rohan, 'Rigidity and the Affliction of Capitalist Property: Colonial Land Revenue and the Recasting of Nature', *Studies in History*, vol. XX, no. 2, n.s., 2004, pp. 237–72.

Das, P.K., *Indian Monsoon*, National Book Trust, Delhi, 1968, rev. edn. 2002.

Dean, Jeffrey S., 'Complexity Theory and Socio-cultural Change in the American Southwest', in Roderick J. McIntosh, Joseph A. Tainter and Susan Keech McIntosh (eds.), *The Way the Wind Blows: Climate, History, and Human Action*, Columbia University Press, New York, pp. 89–118.

Devra, G.S.L., 'Environmental Crisis and Social Dismemberment in Northwest India During the Pre-Colonial Period', paper presented at the *Public Lecture Series*, Nehru Memorial Museum & Library, New Delhi on 4 November 2011, pp. 1-16.

———, 'Desertification and Problem of Delimitation of Rajputana Desert During the Medieval Period', *Human Ecology*, Special Issue, no. 7, 1999, pp. 97-107.

———, 'Physiography and Environment of Thar Desert of Indian Subcontinent During Seventh and Eighth Century', Draft paper presented at the 6th International Conference on 'Perspective on Rajasthan' organized by the Institute of Rajasthan Studies, Jaipur, 6 Januaray 2010.

———, 'Problems in the Delimitation of the Rajputana Desert During the Medieval Period', in Rakesh Hooja and Rajendra Joshi (eds.), *Desert, Drought & Development: Studies in Resource Management and Sustainability*, Rawat Publications, Jaipur, 1999, pp. 371–82.

———, *Rajasthan ki Prashashnik Vyavastha*, Dharti Prakashan, Bikaner, 1981.

———, 'Nature and Incidence of Rokad Rakam (non-agricultural taxes) in the Land Revenue System of the Bikaner State (1650-1700 AD)', *Proceeding of Indian History Congress*, Calcutta, 1976, pp. 190–5.

Dhavalikar, M.K., 'Culture-Environment Interface: A Historical Perspective', Presidential Address, Archaeology, Numismatic and Epigraphy Section, *Indian History Congress*, 52nd Session, Delhi, 1992.

———, 'Green Imperialism: Monsoon in Antiquity and Human Response', *Man and Environment*, vol. XXVI, no. 2, 2001, pp. 17–28.

——, 'The Golden Age and After: Perspectives in Historical Archaeology', *Proceedings of Indian History Congress*, 60th Session, Calicut, 1999, pp. 1–18.

Diamond, Jared, *Guns, Germs and Steel: The Fate of Human Societies*, W.W. Norton, New York, 2005.

Divyabhanusinh, *The End of a Trail: The Cheetah in India*, Oxford University Press, Delhi, 1999.

——, *The Story of Asia's Lion*, Marg Publications, Mumbai, 2005.

Decision Centre for Desert City-II Mission Statement. Point no. 6. Available on http://dcdc.asu.edu/about-us/reports (accessed on 03-04-2012).

Eaton, Richard, *The Rise of Islam and the Bengal Frontier, 1204–1760*, Oxford University Press, Delhi, 1997.

Ellis, Andrew W., Timothy W. Hawkins, Robert C. Balling Jr. and Patricia Gober, 'Estimating Future Runoff Levels for a Semi-arid Fluvial System in Central Arizona, USA', *Climate Research*, vol. XXXV, 2008, pp. 227–39.

Fagan, Brian M., *People of the Earth: An Introduction to World Prehistory*, Scott, Foresman And Company, Glenview, Illinois, Boston, London, 1989.

——, *The Great Warming: Climate Change and the Rise and Fall of Civilizations*, Bloomsbury Press, New York, 2008.

——, *The Long Summer: How Climate Changed Civilisation*, Basic Books, New York, 2005.

Febvre, Lucien, *A Geographical Introduction to History*, tr. E.G. Mountford and J.H. Paxton, Kegan Paul, Trubner & Co., London, 1925.

Fein, Jay S. and Pamela L. Stephens (eds.), *Monsoon*, John Willey & Sons, New York, Chichester, Brisbane, Toronto, Singapore, 1987.

Feldhaus, Anne, *Water and Womanhood: Religious Meanings of Rivers in Maharashtra*, Oxford University Press, New York & Delhi, 1995.

Fischer, David Hackett, 'Climate and History: Priorities for Research', in Robert I. Rotberg and Theodore K. Rabb (eds.), *Climate and History: Studies in Interdisciplinary History*, Princeton University Press, Princeton, New Jersey, 1981, pp. 241–50.

Fisher, R.J., *If Rain Doesn't Come: An Anthropological Study of Drought and Human Ecology in Western Rajasthan*, Manohar, Delhi, 1997.

Friedman, Robert Marc, *Appropriating the Weather: Vilhelm Bkerknes and the Construction of a Modern Meteorology*, Cornell University Press, Ithaca and London, 1989.

Frisinger, H.H., *The History of Meteorology: to 1800*, Science History Publications, New York, 1977.

Gadgil, Madhav and Ramchandra Guha, *This Fissured Land: An Ecological History of India*, Oxford University Press, Delhi, 1992.

Ghosh, A., 'The Rajputana Desert: Its Archaeological Aspects', in S.P. Gupta (ed.), *A. Stein: An Archaeological Tour Along the Ghaggar-Hakra River*, Kusumanjali Prakashan, Meerut, 1989, pp. 98-106.

Gold, Ann Grodzins and Bhoju Ram Gujar, *In the Times of Trees and Sorrows: Natural, Power and Memory in Rajasthan*, Duke University Press, Durham & London, 2002.

———, 'Of Gods Trees & Boundaries', in N.K. Singhi and Rajendra Joshi (eds.), *Folk, Faith & Feudalism*, Rawat Publications, New Delhi, 1995, pp. 33-54.

Golden, Peter B., *An Introduction to the History of the Turkic Peoples: Ethnogenesis and State Formation in Medieval and Early Modern Eurasia and the Middle East*, Otto Harrassowitz, Wiesbarden, 1992.

Grove, Richard and Vinita Damodaran, 'Imperialism, Intellectual Networks, and Environmental Change: Origins and Evolution of Global Environmental History, 1676-2000', Part I: *Economic and Political Weekly*, vol. XLI, no. 41, 14 October 2006, pp. 4345–54; Part II: *Economic and Political Weekly*, vol. XLI, no. 42, 21 October 2006, pp. 4497–505.

Grove, Richard H. and John Chappell, 'El Nino Chronology and the History of Global Crises During the Little Ice Age', in Richard H. Grove and John Chappell (eds.), *El Nino: History and Crisis Studies from Asia-Pacific Region*, The White Horse Press, Massachusetts, 2000, pp. 5–34.

Grove, Richard H., Vinita Damodaran and Satpal Sangwan, 'Introduction', in Richard H. Grove, Vinita Damodaran and Satpal Sangwan (eds.), *Nature and The Orient*, Oxford University Press, Delhi, 1998, pp. 1–17.

Grove, Richard H., Vinita Damodaran and Satpal Sangwan (eds.), *Nature and The Orient*, Oxford University Press, Delhi, 1998.

Grove, Richard, *Green Imperialism: Colonial Expansion, Tropical Island Edens and the Origins of Environmentalism, 1600–1860*, Cambridge University Press, Cambridge, 1995.

Guha, Sumit, *Environment and Ethnicity in India, 1200–1991*, Cambridge University Press, Cambridge, 1999.

———, *Health and Population in South Asia from the Earliest Times to the Present*, Permanent Black, Ranikhet, 2001.

Gupta, Dipankar, 'Hierarchy and Difference: An Introduction', in Dipankar Gupta (ed.), *Social Stratification*, Oxford University Press, New Delhi, 1991, pp. 1–21.

Gupta, G.R., *Marriage, Religion and Society: Patterns of Change in an Indian Village*, Sterling Publishers, Delhi, 1976.

Gupta S.P. and D. Singh, 'Famine in the Territories of Amber (1650–1770)', *Proceedings of Rajasthan History Congress*, Ajmer, 1975, pp. 52–5.

Gupta, S.P., *The Agrarian System of Eastern Rajasthan (c. 1650–1750)*, Manohar, Delhi, 1986.

———, 'Reconstructing the Political and Economic Profile of Rajasthan', *Proceedings of Indian History Congress*, 55th Session, Aligarh, 1994, pp. 159–72.

Habib, Irfan, 'Classifying Pre-Colonial India', in Harbans Mukhia (ed.), *Feudalism and Non-European Societies*, Frank Cass and Company, London, 1985, pp. 44–53.

———, *Medieval Technology Exchanges Between India and the Islamic World*, Viveka Publications, Aligarh, 1985.

———, *The Agrarian System of Mughal India 1556–1707*, 2nd rev. edn., Oxford University Press, Delhi, 2002.

Hannah, David M., Paul J. Wood and Jonathan P. Sadler, 'Ecohydrology and Hydroecology: A "New Paradigm"?' *Hydrological Processes*, vol. XVIII, 2004, pp. 3439–45.

Harrison, Mark, *Climate and Constitutions: Health, Race, Environ-ment and British Imperialism in India, 1600–1850*, Oxford University Press, Delhi, 1999.

Hasan, Saiyid Nurul, *Thoughts on Agrarian Relations in Mughal India*, People's Publishing House, Delhi, 1973.

Hegewald, Julia, 'Water Architecture in Rajasthan', in Giles Tillotson (ed.), *Stones in the Sand: The Architecture of Rajasthan*, Marg Publication, Mumbai, 2001, pp. 78–89.

Honderich, Ted (ed.), *The Oxford Companion to Philosophy*, Oxford University Press, Oxford and New York, 1995.

Hooja, Rima, *A History of Rajasthan*, Rupa & Company, Delhi, 2006.

Hultzsch, E. (ed.), *Corpus Inscriptionum Indicarum*, vol. I, *Inscriptions of Asoka*, Archaeological Survey of India, Delhi, rpt. 1991.

Husain, Iqbal (ed.), *Karl Marx on India*, Tulika Books, Delhi, 2006.

Inden, Ronald, *Imagining India*, Basil Blackwell, Oxford, 1990.

Jain, V.K., *Trade and Traders in Western India*, Munshiram Manoharlal, Delhi, 1990.

Jain-Neubauer, Jutta, *The Step Wells of Gujarat*, Abhinav Prakashan, Delhi, 1981.

Jamindar, Rajesh, 'Agricultural Economy of Gujarat: Aspects and Appraisal up to 1300 AD', in L. Gopal and V.C. Srivastava (eds.), *History of Science, Philosophy and Culture in Indian Civilisation*, vol. V, pt. 1: *History of Agriculture in India (up to c. 1200 AD)*, Concept Publishing, Delhi, 2008, pp. 702-45.

Jeffery, Roger (ed.), *The Social Construction of Indian Forests*, Manohar, Delhi, 1998.

Jha, D.N. (ed.), *Feudal Social Formations in Early India*, Chanakya Publications, Delhi, 1987.

Jodha, N.S., *Life on the Edge: Sustaining Agriculture and Community Resources in Fragile Environments*, Oxford University Press, Delhi, 2001.

Joshi, Govind Lal, 'Harauti Ki Krishi Kahavaten', *Maru Bharati*, vol. XIII, 1965–6, pp. 43-58.

Kapur, Nandini Sinha, *State Formation in Rajasthan: Mewar during the Seventh-Fifteenth Centuries*, Delhi, Manohar, 2002.

Karl Marx: Articles on India, Peoples' Publishing House, Bombay, 1943.

Kavoori, Purnendu, *Pastoralism in Expansion: The Transhuming Herders of Western Rajasthan*, Oxford University Press, Delhi, 1999.

Khan, I.A. and Ravindra Kumar, 'The Mansagar Dam of Amber', in Anirudha Ray and S.K. Bagchi (eds.), *Ancient and Medieval Technologies in India*, Sundeep Prakashan, Delhi, 1986, pp. 25–40.

Khan, Mohd. Afzal, 'Environment and Pollution in Mughal India, *Islamic Culture*, vol. LXXVI, no. 1, January 2002, pp. 101–16.

Khazanov, Anatoly M., *Nomads and the Outside World*, University of Wisconsin Press, Madison, 1994.

Kochhar, Rajesh, *The Vedic People, Their History and Geography*, Orient Longman, Hyderabad, 2000.

Kolff, Dirk H.A., *Naukar, Rajputs and Sepoy: The Ethnohistory of the Military Labour Market in Hindustan, 1450-1850*, Cambridge University Press, Cambridge, 1990.

Kosambi, D.D., *An Introduction to the Study of Indian History*, Popular Book Depot, Bombay, 1975.

Kumar, K.S. Kavi, 'Climate Sensitivity of Indian Agriculture', Working Paper 43/2009, Madras School of Economics, Chennai, 2009.

Kumar, Mayank, 'Claims on Natural Resources: Exploring the Role of Political Power in Pre-Colonial Rajasthan, India', *Conservation and Society*, vol. III, no. 1, June 2005, pp. 134–49.

——, 'Ecology and Traditional Systems of Water Management: Revisiting Medieval Rajasthan', in Mahesh Rangarajan (ed.), *Environmental Issues in India*, Pearson & Longman, Delhi, 2006, pp. 70–96.

——, 'Flexibility and Adaptability: Agrarian Expansion and Traditions of Water Management', *The ICFAI University Journal of History and Culture*, vol. III, no. 3, 2009, pp. 31–55.

——, 'Situating the Environment: Settlement, Irrigation, and Agriculture in Pre-colonial Rajasthan', *Studies in History*, vol. XXIV, no. 2, n.s., 2008, pp. 211–33.

——, 'Perceiving Landscape: Regional Identity Formations in Early Modern India', *History Today: Journal of History and Historical Archaeology*, no. 9, 2008, pp. 57-61.

——, 'Nature, Social Stratification and Marginalization in Rajasthan, 1650-1850 CE', paper accepted for publication at Nehru Memorial Museum & Library, Delhi as part of Occasional Paper Series.

Ladurie, E. Le Roy, *The Territory of Historian*, tr. Sian Reynolds and Ben Reynolds, The Harvester Press, Hassocks, 1979.

——, *Times of Feast, Times of Famine: A History of Climate Since the Year 1000*, tr. Barbara Bray, Doubleday & Company Inc, New York, 1971.

Lakshmi, I., 'Wet Crop Cultivation, the Brahamans and the Cultivating Communities in Medieval Telingana', *Indian Historical Review*, vol. XXXI, nos. 1 & 2, January–July 2004, pp. 60–95.

Laḷas, Sita Ram, *Rajasthani Hindi Sabad Kosh*, vol. I, Rajasthani Shodh Sansthan, Jodhpur, 1986.

Lamb, H.H., 'An Approach to the Study of the Development of Climate and its Impact in Human Affairs', in T.M.L. Wigley, M.J. Ingram and G. Farmer (eds.), *Climate and History: Studies in Past Climate and Their Impact on Man*, Cambridge University Press, Cambridge, 1981, pp. 289–90.

——, *Climate, History and the Modern World*, Methuen, London and New York, 1982.

Lavee, Hanoch, Jean Posen and Aaron Yair, 'Evidence of High Efficiency Water-Harvesting by Ancient Farmers in the Negev Desert, Israel', *Journal of Arid Environments*, vol. XXXV, no. 2, February 1997, pp. 341–8.

Livingstone, Morna, *Steps to Water: The Ancient Stepwells of India*, Princeton Architecture Press, New York, 2002.

Lodrick, Deryck O., 'Rajasthan As a Region: Myth or Reality?' in Karine Schomer, Joan L. Erdman, Deryck O. Lodrick and Lloyd I. Rudolph (eds.), *The Idea of Rajasthan*, vol. I, Manohar, Delhi, 1994, pp. 1-44.

Ludden, David, 'Ecological Zones and the Cultural Economy of Irrigation in Southern Tamilnadu', *South Asia*, vol. I, no. 1, 1978, pp. 1-13.

———, *An Agrarian History of South Asia: The New Cambridge History of India*, vol. IV.4, Cambridge University Press, Cambridge, 1999.

MacKenzie, John M., *The Empire of Nature: Hunting, Conservation and British Imperialism*, Manchester University Press, New York, 1988.

Madella, Marco and Dorian Q. Fuller, 'Palaeoecology and the Harappan Civilisation of South Asia: A Reconsideration', *Quaternary Science Reviews*, vol. XXV, 2006, pp. 1283–1301.

Maheswari, Hira Lal, *Jambhoji, Bishnoi Sampradaya aur Sahitya*, B.R. Publications, Calcutta, 1970.

Malhotra, S.P., 'Man and the Desert', in Rakesh Hooja and Rajendra Joshi (eds.), *Desert Draught and Development*, Rawat Publishers, Jaipur, 1999, pp. 17-33.

Mallah, Qasid Hussain, 'An Archaeological Assessment of Taung Valley of Sindh-Kohistan Pakistan', *Ancient Asia*, vol. II, 2009–10, pp. 93–112.

Margbandhu, C., 'A Late Medieval Step-well at Nimrana, District-Alwar', in P. Chennareddy (ed.), *An Archaeological and Cultural Study: Researches in Archaeology, History and Culture in the New Millennium*, Sharada Publishing House, Delhi, pp. 49-57.

McIntosh, Jane R, *The Ancient Indus Valley: New Perspectives*, ABC-Clio, Inc, Santa Barbara, 2008.

McIntosh, Roderick J., Joseph A. Tainter and Susan Keech McIntosh (eds.), *The Way the Wind Blows: Climate, History, and Human Action*, Columbia University Press, New York, 2000.

Milly, P.C.D., Julio Betancourt, Malin Falkenmark, Robert M. Hirsch, Zbigniew W. Kundzewicz, Dennis Lettenmaier and Ronald J. Stouffer, 'Stationarity Is Dead: Whither Water Management', *Science*, 319, 2008, pp. 573-4.

Misra, Anupam, *Rajasthan ki Rajat Boonden*, Gandhi Shanti Pratishthan, Delhi, 1995.

Misra, V.N., 'Climate, a Factor in the Rise and Fall of the Indus Civilization: Evidence from Rajasthan and Beyond', in B.B. Lal and S.P. Gupta (eds.), *Frontiers of the Indus Civilizations*, Books and Books, New Delhi, 1984, pp. 461-89.

Monier-Williams Monier, *Sanskrit-English Dictionary: Etymologically and Philologically Arranged with Special Reference to Cognate Indo-European Languages*, Clarendon Press, Oxford, 1951.

Montgomery, David R., *Dirt: The Erosion of Civilizations*, University of California Press, Berkeley, 2007.

Moosvi, Shireen, 'Ecology, Population Distribution and Settlement Pattern in Mughal India,' *Man and Environment*, vol. XIV, no. 1, 1989, pp. 109–16.

Moreland, W.H., *India at the Death of Akbar: An Economic Study*, Reprints and Trans Publications, Delhi, 1974.

Morrison, Kathleen D., 'Environmental History, the Spice Trade, and the State in South India', in Mahesh Rangarajan and K. Sivaramakrishnan, (eds.), *India's Environmental History*, vol. I, *From Ancient Times to the Colonial Period*, Permanent Black, Ranikhet, 2012, pp. 296–326.

——, 'Production and Landscape in the Vijayanagara Metropolitan Region: Contributions of the Vijayanagara Metropolitan Survey', in J.M. Fritz, T. Raczek, and R. Brubaker (eds.), *Vijayanagara: Archaeological Exploration 1990-2000: Papers in Memory of Channabasappa S. Patil*, Vijayanagara Research Project Monographs, vol. X, Manohar and AIIS, 2005, pp. 423-34.

——, *Fields of Victory: Vijayanagara and the Course of Intensification*, Munshiram Manoharlal, Delhi, 2000.

Morrison, Kathleen D and L.L. Junker (eds.), *Forager-Traders in South and Southeast Asia: Long-Term Histories*, Cambridge University Press, Cambridge, 2002.

Mosse, David, *Rule of Water; Statecraft, Ecology and Collective Action in South India*, Oxford University Press, Delhi, 2003.

Mukhia, Harbans, 'Was there Feudalism in Indian History?', *Proceedings of Indian History Congress*, 40th Session, Waltair, 1979, pp. 229–80.

Murthy, M.L.K., 'Environment, Royal Policy and Social Formations in the Eastern Ghats, South India', *Proceedings* of *Indian History Congress*, Delhi, 1993, pp. 615-31.

Nagrath, Pooja, 'A Comparative Analysis of Rainwater Harvesting Practices Implemented in Arizona and Rajasthan', unpublished M.Sc. thesis, Arizona State University, Tempe, 2008.

Nahata, Agarchand, ed., 'Desh Varnan', *Maru Bharti*, vol. II, no. 2, 1954, pp. 49–51.

——, 'Thali Varnan', *Maru Bharti*, vol. II, no. 1, 1954, pp. 81–3.

———, 'Rajasthani Gadh Main Varsha Varnan', *Maru Bharti*, vols. 1–2, 1952–5, pp. 63–4.

Nandi, R.N., *Ideology and Environment: Situating the Origins of Vedic Culture*, Aakar Books, Delhi, 2009.

Nath, Pratyaya, 'Rethinking Early Mughal Warfare: Babur's Pitched Battles, 1499-1529', in Raziuddin Aquil and Kaushik Roy (eds.), *Warfare, Religion and Society in Indian History*, Manohar, Delhi, 2012, pp. 109–45.

———, 'From Mocking to Becoming the 'Lords of the Elephants', War-Elephants and the Mughals, 1526–c. 1590', paper presented at a Workshop on Environments and Histories, organized by Nehru Memorial Museum & Library, 12 May 2012.

O'Brien, A.G., *The Ancient Chronology of Thar: The Bhattika, Laulika and Sindh Eras*, Oxford University Press, Delhi, 1996.

Ojha, G.H., *Rajputane ka Pracheen Itihas*, Rajasthani Granthagar, Jodhpur (1st pub. in 1926), 3rd rev. edn., 2009.

Oldham, Surgeon-Major C.F., 'The Saraswati and the Lost River of the Indian Desert,' *Journal of Royal Society*, 1893, pp. 49–76.

Pandey, Deep Narayan, Anil K. Gupta and David M. Anderson, 'Rainwater Harvesting as an Adaptation to Climate Change', *Current Science*, vol. LXXXV, no. 1, 10 July 2003, pp. 46–59.

Pandian, Anand S, 'Predatory Care: The Imperial Hunt in the Mughal and British India', *Journal of Historical Sociology*, vol. XIV, no. 1, March 2001, pp. 79-107.

Parasher-Sen, Aloka, 'Of Tribes, Hunters and Barbarians: Forest Dwellers in the Mauryan Period', *Studies in History*, vol. XIV, no. 2, n.s., 1998, pp. 173–91.

Patel, Prabhubhai K., 'Construction and Structure of Step Wells', *Puratattva*, vol. XXVIII, 1997-8, pp. 91-103.

Pelsaert, Francisco, *Remonstrantie*, tr. W.H. Moreland and Pieter Geyl, *Jahangir's India*, W. Heffer and Sons, Cambridge, 1925.

Pemaram, *The Religious Movements in Medieval Rajasthan*, Archana Prakashan, Ajmer, 1977.

Psillos, Stathis and Martin Curd, 'Introduction,' in Stathis Psillos and Martin Curd (eds.), *The Routledge Companion to Philosophy of Science*, Routledge, Abingdon and New York, 2008, pp. 1–9.

Rabb, Theodore K., 'The Historian and the Climatologist', in Robert I. Rotberg and Theodore K. Rabb (eds.), *Climate and History: Studies in Interdisciplinary History*, Princeton University Press, Princeton, New Jersey, 1981, pp. 251–7.

Rad, Ulrich Von et al., 'A 5000-yr Record of Climate Change in Varved Sediments from the Oxygen Minimum Zone off Pakistan, Northeastern Arabian Sea', *Quaternary Research*, vol. LI, 1999, pp. 39–53.

Radhakrishna, B.P. and S.S. Merh (eds.), *Vedic Sarasvati: Evolutionary History of a Lost River of Northwestern India*, Memoir 42, Geological Survey of India, Bangalore, 1999.

Rahman, A, *Maharaja Sawai Jai Singh II and Indian Renaissance*, Navrang, Delhi, 1987.

Rajalakshmi, T.K., 'The Riddle of a River'. *Frontline*, vol. XIX, Issue 16, 3-16 August 2002. See http://www.thehindu.com/fline/fl1916/19160970.htm (accessed on 9 March 2010).

Ramaswamy, C., 'Monsoon Over the Indus Valley during the Harappan Period', *Nature*, no. 217, pp. 628–9.

Ramusack, Barbara N., *The Indian Princes and their States*, 1st Asian edition, Cambridge University Press, New York, 2005.

Rangarajan, Mahesh and K Sivaramakrishnan (eds.), *India's Environmental History*, vol. I: *From Ancient Times to the Colonial Period*, Permanent Black, Ranikhet, 2012.

——, *India's Environmental History*, vol. II: *Colonialism, Modernity and Nation*, Permanent Black, Ranikhet, 2012.

Ratnagar, Shereen, 'Pastoralism as an Issue in Historical Research', *Studies in History*, vol. VII, no. 2, n.s., 1991, pp. 181-93.

——, *Understanding Harappa: Civilization in the Greater Indus Valley*, Tulika, Delhi, 2001.

Raverty, Major H.G., *The Mihran of Sind and its Tributaries*, Sang-E-Meel Publications, Lahore, 1979.

Ray, R., *Ancient Settlement Patterns in Eastern India: Prehistoric Period*, Pearl Publishers, Delhi, 1987.

Ray, Rajat Kant, 'Colonel James Tod, Muhanota Nainasi and the Rajputs', *Indian Historical Review*, vol. XXV, no. 2, June 1999, pp. 100–11.

Rezavi, S. Ali Nadeem, 'Kuldhara in Jaisalmer State: Social and Economic Implications of the remains of Medieval Settlement', *Proceedings of Indian History Congress*, 56th Session, Rabindra Bharti, Calcutta, 1995, pp. 312–38.

Rosen, Arlene Miller, *Social Responses to Climate Changes in the Ancient Near East*, AltaMira Press, Maryland, 2007.

Rosenfeld, J., *Eye of the Storm: Inside the World's Deadliest Hurricanes, Tornadoes and Blizzards*, Plenum Trade, New York and London, 1999.

Rosin, R.T., *Land Reforms and Agrarian Change: Study of a Marwar Village from Raj to Swaraj*, Rawat, Jaipur, 1987.

Roy, Tirthankar, *Natural Disasters and Indian History*, Oxford University Press, Delhi, 2012.

Sahai, Nandita Prasad, *Politics of Patronage and Protest: The State, Society, and Artisans in Early Modern Rajasthan*, Oxford University Press, Delhi, 2006.

Sahal, Kanhiya Lal, 'Rajasthan ki Varsha Sambandhi Kahavaten', *Maru Bharati*, vol. IV, 1956, pp. 8–18.

Sahay, Neeraj, 'Environment, Settlement History and the Emer-gence of An Agrarian Region in Early Medieval Andhra', unpublished Ph.D thesis, University of Delhi, 2009.

Scott, James C., *Seeing Like a State: How Certain Schemes to Improve the Human Condition Have Failed*, Yale University Press, New Haven, 1998.

——, *Weapons of the Weak: Everyday Forms of Peasant Resistance*, Yale University Press, New Haven and London, 1985.

Sen, Surendra Nath (ed.), *Indian Travels of Thevenot and Careri*, National Archives of India, New Delhi, 1949.

Sharma D., *Rajasthan Through the Ages*, Rajasthan State Archives, Bikaner, 1966.

——, 'Some dates in the Bhattic Era and its Connected Problems', *Indian Historical Quarterly*, vol. XXXV, 1959, pp. 227–39.

Sharma, G.D., *Rajput Polity: A Study of Politics and Administration of the State of Marwar, 1638–1749*, Manohar, Delhi, 1977.

Sharma, G.N., *Social Life in Medieval Rajasthan*, Lakshmi Narain Agarwal Educational Publishers, Agra, 1968.

——, *Rajasthan ke Itihas ke Srota*, pt. 1, Rajasthan Hindi Grantha Akadami, Jaipur, 1983.

Sharma, Jai Shankar Dev Shankar, 'Prakriti se Varsha Gyan ki Banagi', *Rajasthan Bharati*, vol. VI, nos. 3–4, 1956, pp. 81–3.

Sharma, R.S., *Economic History of Early India*, Viva Books, Delhi, 2001.

——, *Indian Feudalism: c. 300–1200*, Calcutta, 1965, 2nd edn., Macmillan India, Delhi, 1980.

——, *Social Change in Early Medieval India (c. AD 500–1200)*, Peoples' Publishing House, Delhi, 1969.

——, *Urban Decay in India (c.300–c.1000)*, Munshiram Manoharlal, Delhi, 1987.

Sharma, Rajiv and Syed Ali Nadeem Rezavi, 'Aspects of Hydraulic Engineering in Medieval Rajasthan: A Case Study of Water System in Jaigarh Fort', in Ahsan Jan Qaisar and Som Prakash Verma (eds.), *Art and Culture*, Publications Scheme, Jaipur, 1993, pp. 129–31.

Sharma, Yogesh, 'The Circuit of Life: Water and Water Reservoirs in Pre-modern India', *Studies in History*, vol. XXV, no. 1, n.s., January-June 2009, pp. 69–108.

Shelat, Bharati, 'The Step-well Inscription of Gujarat: A Cultural Study', in P. Chennareddy (ed.), *'Krishnavinandan' Archaeological, Historical and Cultural Studies* (Festschriftt of V. V. Krisnasastry), Research Press, Delhi, pp. 173–85.

Siddle, David, 'Goats, Marginality and the 'Dangerous Other', *Environment and History*, vol. XV, no. 4, November 2009, pp. 521–36.

Singh, Chetan, *Natural Premises: Ecology and Peasant Life in the Western Himalaya 1800–1950*, Oxford University Press, Delhi, 1998.

———, *Region and Empire: Punjab in the Seventeenth Century*, Oxford University Press, Delhi, 1991.

Singh, Dilbagh, 'The Role of Mahajans in the Rural Economy in Eastern Rajasthan During the 18th Century', *Social Scientist*, vol. II, no. 10, October 1974, pp. 22–31

———, 'Regulating the Domestic: Notes on the Pre-colonial State and the Family', *Studies in History*, vol. 19, no. 1, n.s., 2003, pp. 69–86.

———, *The State, Landlords and Peasants*, Manohar, Delhi, 1990.

Singh, Gurdeep, 'The Indus Valley Culture (Seen in the Context of Post-Glacial Climate and Ecological Studies in North-West India)', *Archaeology and Physical Anthropology in Oceania*, vol. VI, no. 2, pp. 177–89.

Singh, Mahavir Gahlot (ed.), *Dhola Maru ra Duha*, Rajasthani Granthagar, Jodhpur, 1985.

Singh, Shashi Bhushan, *Sarnath ki Baudh Murtikala*, Swati Publications, Delhi, 2009.

Singh, Upinder, *A History of Ancient and Early Medieval India: From the Stone Age to the 12th* Century, Pearson Longman, Delhi, 2008.

Singh, Virendra, *Jaisalmer Rajgharane ka Kendriya Satta se Sambandh*, Rajasthani Granthagar, Jodhpur, 2011.

Sivaramakrishnan, K. and Ismael Vaccaro, 'Introduction. Post-industrial Natures: Hyper-mobility and Place-attachments', *Social Anthropology*, vol. XIV, no. 3, 2006, pp. 301–17.

Smith, Vincent A., *Oxford History of India* (1919), 3rd edn., Revised by Mortimer Wheeler, A.L. Basham and T.G. Percival Spear, Oxford University Press, Oxford, 1958.

Sontheimer, Gunther-Dietz, 'God as King For All: The Sanskrit Mallari Mahatmya and its Context', in Anne Feldhaus, Aditya Malik and Heidrun Bruckner (eds.), *King of Hunter, Warriors and Shepherds*, Manohar, Delhi, 1997, pp. 208-35.

——, *Pastoral Deities in Western India*, tr. Anne Feldhaus, Oxford University Press, New York, 1989.

Spooner, Brian, '"Baluchistan": Geography, History and Ethnography' in Ehsan Yarshater, (General Editor), *The Encyclopædia Iranica*, Center for Iranian Studies, Columbia University New York, http://www.iranicaonline.org/articles/baluchistan-i (accessed on 18 September 2011).

Srivastava, Vinay Kumar, 'The Rathore Rajput Heroes of Rajasthan: Some Reflections on John Smith's *The Epic of Pabuji*', *Modern Asian Studies*, vol. XXVIII, no. 3, 1994, pp. 597–9.

Stein, Burton, *Peasant, State and Society in Medieval South India*, Oxford University Press, Delhi, 1980.

——, *The New Cambridge History of India*, vol. II: *Vijayanagara*, Cambridge University Press, Cambridge, 1994.

Step Wells of Rajasthan, Department of Tourism Art and Culture, Government of Rajasthan, Jaipur, 1998.

Talbot, Cynthia, *Pre-Colonial India in Practice: Society, Region and Identity in Medieval Andhra*, Oxford University Press, Delhi, 2001.

Taub, Liba, *Ancient Meteorology*, Routledge, London and New York, 2003.

Trautmann, Thomas R., *The Clash of Chronologies: Ancient India in the Modern World*, Yoda Press, Delhi, 2009.

Vasavi, A.R., *Harbingers of Rain: Land & life in South India*, Oxford University Press, Delhi, 1999.

Vashishtha, Neelima, 'A Note on the Sculptures of the Raj Samudra Lake in Rajasthan', in Ahsan Jan Qaisar and Som Prakash Verma (eds.), *Art and Culture*, Publications Scheme, Jaipur, 1993, pp. 111–18.

Venke, Robert J., *Patterns in Prehistory: Humankind's First Three Million Years*, Oxford University Press, New York & Oxford, 4th edn., 1999.

Verma, H.C., *Harvesting Water and Rationalization of Agriculture in North Medieval India: Thirteenth-Sixteenth Centuries*, Anamika, Delhi, 2001.

Wahi, Tripta, 'Water Resources and Agricultural Landscape: Precolonial Punjab', in Indu Banga (ed.), *Five Punjabi Centuries: Polity, Economy, Society and Culture, c.1500-1990*, Manohar, Delhi, 1997, pp. 267–84.

Warren, B.A., 'Ancient and Medieval Records of the Monsoon Winds and Currents of the Indian Ocean', in Jay S. Fein and Pamela L. Stephens (eds.), *Monsoon*, John Wiley & Sons, New York, Chichester, Brisbane, Toronto, Singapore, 1987, pp. 137–58.

Watters, T., *On Yuan Chwang Travels in India*, vol. II, Munshiram Manoharlal, Delhi, 1973.

Welch, Anthony, 'Hydraulic Architecture in Medieval India: The Tughluqs', *Environmental Design*, vol. II, 1985, pp. 74–81.

Yadav, R.K., 'Changes in the Large-scale Features Associated with the Indian Summer Monsoon in the Recent Decades', *International Journal of Climatology*, no. 29, 2009, pp. 117–33.

Ziegler, Norman P., 'Evolution of the Rathor State of Marvar: Horses, Structural Change and Warfare, in Karine Schomer, Joan L. Erdman, Deryck O Lodrick, Lloyd L Rudolph (eds.), *The Idea of Rajasthan, Explorations in Regional Identity*, vol. II, Manohar, Delhi, 1994, pp. 192-216.

Zimmerman, F., 'Monsoon in Traditional Culture', in Jay S. Fein and Pamela L. Stephens (eds.), *Monsoon*, John Willey & Sons, New York, Chichester, Brisbane, Toronto, Singapore, 1987, pp. 51-76.

——, *The Jungle & the Aroma of Meats: Ecological Themes in Hindu Medicine*, University of California Press, Berkeley, Los Angeles, London, 1987.

Index